Salt, Lemons, Vinegar, and Baking Soda

Salt, Lemons, Vinegar, and Baking Soda

HUNDREDS OF EARTH-FRIENDLY HOUSEHOLD PROJECTS, SOLUTIONS, AND FORMULAS

Shea Zukowski

METRO BOOKS
NEW YORK

A QUIRK PACKAGING BOOK

This 2009 edition published by Metro Books,
by arrangement with Quirk Packaging, Inc.

Design by Lynne Yeamans and Nancy Leonard

Metro Books
122 Fifth Avenue
New York, NY 10011

ISBN: 978-1-4351-1697-9

Printed and bound in China

1 3 5 7 9 10 8 6 4 2

Acknowledgments

Unlike the hundreds of formulas you'll find in this book, each requiring only a few simple ingredients to work their magic, putting them together into a book is a far more complex endeavor that requires the talents of many. To that end, I'd like to extend my utmost thanks to Erica Heisman and Peter Norton at Barnes and Noble, Inc., for helping make the world a little greener by recognizing the need for a book like this.

Likewise, I am grateful for the pleasure of collaborating with the brilliant team at Quirk Packaging, including Sharyn Rosart for her enthusiastic vision of how this project should take shape; my editor, Erin Canning, for her great ideas and keen attention to detail; Lynne Yeamans and Nancy Leonard, whose inspired design skills were delightful at every turn; and Deri Reed, whose copyediting talents once again saved the day.

Finally, my most heartfelt appreciation goes to my family, Stan, Isaiah and Eli, whom I can always count on to share a few messes that need a little cleaning, and plenty of laughter along the way.

Contents

Introduction

These days there are hundreds of products on the market claiming to be "green"—a natural response to consumer demand for more earth-friendly options and for a cleaner, less toxic home environment. But you may be pleasantly surprised to learn that "going green" doesn't have to mean spending a lot of money. You most likely already have four natural, everyday items in your home that can accomplish hundreds of household tasks. So forego buying a dozen different products and stock up on the Fab Four!

Meet The Fab Four: Salt, Lemons, Vinegar, and Baking Soda. The Fab Four are the all-you-need all-stars for kitchen, laundry, personal, household, outdoor, and pet care tasks. Each one is inexpensive and earth-friendly, and provides a wealth of uses, whether on its own or in combination with another Fab Four ingredient or other basic product. A brief introduction to each of these brilliant, versatile substances follows; turn to the individual chapters for a more detailed look.

✦ ✦ ✦

SALT

When most of us think of salt, our thoughts naturally turn to food. After all, saltiness is one of the five basic tastes (the other four are sweetness, bitterness, sourness, and umami). And for thousands of years people have relied on salt for survival, as it offers a highly effective means of preserving food. In fact, its value was once so precious and its availability in some regions so scarce that it was actively traded in many early societies and even considered currency; in some cultures salt was traded for gold. NOTE: For the purposes of this book, use table salt unless a formula directs otherwise.

LEMONS

Compared to salt there are fewer types of lemons to think about (most lemons sold today are generally the same), but the list of ways to use this amazing fruit is just as long. Lemons have mild bleaching properties that make them ideal for stain removal on clothing and fabrics. Used in combination with salt or baking soda, they offer a cleaning boost that rivals commercial products. Plus, lemons can be used as part of a low-cost beauty regime to lighten hair or age spots. And best of all, lemons impart a signature scent that's fresh and invigorating— and all natural!

♦ ♦ ♦

VINEGAR

Vinegar is truly amazing because its applications are just as varied as its sources. Basically, any food containing sugar or starch that can be converted into alcohol can be made into vinegar. As a cleaning agent, vinegar possesses germ-fighting abilities and will leave practically every surface in the kitchen and bath sanitized and streak-free; as a laundry aid, it is legendary for its power to soften fabrics and remove stains; and outdoors, vinegar can be a gardener's best friend. NOTE: For the purposes of this book, use distilled white vinegar unless a formula directs otherwise.

BAKING SODA

Baking soda is not just for sitting in the refrigerator; this is a little box with a whole lot of uses around the house. In the laundry room it boosts the effectiveness of regular detergent and can help lift stains before they've set. Baking soda also has an amazing ability to absorb unpleasant odors wherever they crop up (closets, well-worn shoes, old books, garbage pails, and even a wet dog). And because baking soda is mildly alkaline, it provides a safe and handy way to stop a grease fire or neutralize battery acid. Now that's versatility!

What Else Will You Need?

While all of the formulas in this book revolve around the Fab Four, there are a few other items you should make sure to have on hand.

Brushes and sponges: Keep a good supply of brushes in a variety of sizes. At the very least, you should have a small soft-bristle brush, like a nail brush, and one all-purpose brush. Where sponges are concerned, it's good to have a few with a nylon scrubbing surface attached, as well as a large, soft sponge that can hold a lot of water and is suitable for washing large surfaces.

Buckets and clean rags: For large cleanups, keep a 2-gallon (7.5-liter) plastic bucket on hand. And if you really want to be green, remember to use disposable paper towels for only the most unpleasant types of messes; instead, keep a supply of old cotton socks, rags, and towels on hand. They'll serve you well in most situations.

Cornstarch: As the name implies, cornstarch is essentially the starchy component of the corn grain. Its silky texture makes it an ideal anti-caking agent when mixed with dry ingredients. Among its many useful properties, thickening and absorbing are most commonly utilized.

Cream of tartar: A byproduct of winemaking, cream of tartar is a white powder used in cooking for its mild acidic properties and is essential when transforming baking soda into baking powder. It is also helpful for spot cleaning leather.

Dish soap: Many of the cleaning formulas require just a few drops of dish soap to round them out. For best results, look for an earth-friendly brand that specifically offers grease-fighting abilities.

Essential oils: Some cleaning and personal care formulas call for essential oils to impart a pleasant fragrance as well as therapeutic properties. In most cases, their inclusion is optional, but if you enjoy scents, you might like to stock up on a few. Store essential oils in a cool, dark place and never apply directly to the skin as they are very concentrated.

Hydrogen peroxide: Another useful compound. Added bounus: If you are a dog owner, make sure to always have an unopened bottle of hydrogen peroxide tucked away for skunk emergencies and for cleaning your dog's ears.

Isopropyl alcohol: You most likely have a bottle of rubbing (isopropyl) alcohol in your medicine cabinet. For an effective household disinfectant, use the plain variety without any added coloring and fragrance.

continued on page 18

What Else Will You Need?

Laundry detergent: While most of the formulas are designed to let you launder as usual, it's a good idea to keep fragrance-free and color-safe versions of your favorite detergents around the house. Color-safe products do not contain chlorine bleach.

Linseed oil: Made from the dried, ripe seeds of the flax plant, linseed oil is commonly used with oil paints, as a wood finish, and for the purpose of this book, as a leather treatment. NOTE: Rags soaked in linseed oil are considered a fire hazard, so immediately soak them in water after use and never put them in the dryer.

Olive oil: In the culinary world, olive oil is the natural counterpart to vinegar, with just as many nuances and expensive price tags. For general household use, olive oil is useful as a polisher of wood and leather. An inexpensive version will work fine, but just make sure it's one hundred percent olive oil.

Storage containers: Depending on the size of your project or task, you may be able to get many uses out of some of the formulas in this book. Lidded plastic containers are fine for most types of storage, though for sprays, of course, you should rely on spray bottles that can be purchased in the beauty section of drugstores and department stores. Make sure to clearly label any formula you decide to keep with its ingredients, as well as the date you made it, and store out of reach of children and pets.

The following items are each used in at least one cleaning formula:

Almond Oil

Aluminum foil

Ammonia

Cheesecloth

Club soda

Coffee grounds

Cornmeal

Cotton balls

Food coloring

Parchment paper

Rose water

Sugar

Vegetable oil

Vodka

Before You Begin

While the time-honored formulas in this book are generally safe and effective for a variety of uses, there are a few guidelines to keep in mind:

Don't take unnecessary risks. Sure, you're working with basic ingredients that are relatively benign on their own, but it's simply impossible to predict exactly how each formula will perform with every type of material. So think carefully before you decide to try a formula on a favorite personal item, and consider sending irreplaceable items, such as antiques, out for a professional cleaning.

Test on an inconspicuous place. On the other hand, many of the natural formulas described in this book are tried-and-true, so they most likely are the sorts of cleaning substances your grandmother would have used when she first acquired that keepsake you now love. For best results, always test a formula on an inconspicuous spot to see how it works before applying it to the entire area.

Follow the directions. While it's acceptable to adjust ratios in these formulas (i.e., a stronger solution for an especially tough stain), don't add other ingredients to the mix in an attempt to boost the effectiveness of a formula. The combination of vinegar and bleach, for example, can produce toxic fumes.

Use plastic or glass containers. Unless the directions state otherwise, make it a general rule to mix up your cleaning formulas in plastic or glass containers (some metals will react with the cleaning formulas).

Consult your doctor. If you have sensitive skin or if you're being treated for a medical condition, talk with your doctor before using the personal care remedies in this book. This book is not intended to replace the advice of a physician.

Work quickly and rub lightly. Where stain removal is concerned, these are the two golden rules to keep in mind at all times. The longer a stain sits, the more likely it is to set. And though your inclination may be to scrub vigorously to get it out, a lighter hand is always best. Intense scrubbing can weaken the surrounding material and quite possibly drive the stain deeper into the fabric.

CHAPTER I

Salt

✦ ✦ ✦

Salt is a natural mineral composed of two elements, sodium and chloride, which, when chemically united, form the compound sodium chloride, or NaCl. Salt is abundant in our world and necessary to the healthy functioning of our bodies.

The salt that we use is drawn from two main sources: seawater, which contains a seemingly infinite supply, and solid underground deposits left over when ancient waters evaporated. Salt is extracted from seawater by boiling or evaporation. Deposits of solid, or rock salt (called halite), are mined. Salt has even been found in meteors and on Mars.

Salt varies in color from colorless, when pure, to white, gray, pink, or brown, when solid. Salt dissolves readily in water. Salt crystals can be grown in various sizes and salt companies prepare particles in a wide variety of sizes to meet customer needs. More than 200 million tons of salt are used each year, with literally thousands of uses from personal to industrial.

A Brief History of Salt

So valuable has salt been in human history, that wars were waged over it: To control a country's salt supply was, in effect, to control the fate of its people. Without proper access to salt, large populations were rendered more vulnerable to disease and hardship. And because crops cannot grow in salinated soil, salt was even used in warfare as a long-lasting weapon; in 250 BC, the Carthaginians took on the Greeks and Romans to control the Mediterranean salt supply, but when they were defeated, their lands were rendered infertile when large amounts of salt were plowed into their fields.

Salt was harvested as early as 6000 BC in northern China, and there is evidence that the ancient Egyptians evaporated seawater to obtain salt that they used to preserve meat and fish—and mummies. The ancient Etruscans, the early Romans, the Carthaginians, and the Celts also used evaporation. Many of the earliest trade routes developed around the salt trade, while cities grew near salt centers. Historically, salt was mainly used to preserve and flavor food, as well as in making leather, textiles, and pottery. After the era of industrialization, however, salt became significant in manufacturing, and today the industry claims more than 14,000 uses for salt.

How Salt Works

In the average household, salt's uses tend to revolve around food and cleaning. There are two major culprits responsible for food spoiling: bacteria and enzymes. Salt is able to tackle both head on, and, thus, help preserve food. Bacteria are unable to feed and reproduce in a salty environment, and the presence of salt in food can also delay certain enzyme reactions from occurring.

Around the house, salt has several properties that make it a wonder cleaner with many fantastic uses. When salt is used in a "dry," or undissolved state, it can be mildly abrasive, meaning it can be used to scrub pots and pans (without scratching). It also has some absorbent properties where spills are concerned. And when salt is dissolved in water it becomes a saline solution, capable of helping lift a variety of protein stains, including blood and perspiration, from fibrous materials like clothing and carpets.

When salt is dissolved in water it makes it harder for the water molecules to move around, thus raising the boiling point slightly since it takes more energy for the water to turn into steam; likewise, salt can disrupt water's ability to form crystals, thereby lowering the point at which water can turn to ice and preventing driveways and sidewalks from becoming slick.

A Field Guide to Major Salt Types

REFINED SALT: Almost entirely made up of sodium and chloride, refined salt is most commonly used. Raw salt is refined by going through a purification process that strips it of trace minerals and changes its chemical structure from large crystals into smaller granules

Table salt: Sometimes known as iodized salt, this most common type of refined salt often contains an anti-caking agent that inhibits lumping in humidity and is often fortified with iodine to prevent thyroid disease. NOTE: Table salt should be used in all of the formulas in this book unless another type is specified.

Kosher salt: A refined, coarser-grained salt with no additives (i.e., anti-caking agents). Known for its large, irregularly shaped flakes (the product of a compaction process), which allows the salt to easily draw blood when applied to butchered meat, kosher salt is saltier in taste and is preferable to table salt in many types of food preparation. The structure of the grains allows them to dissolve easily and provides a smoother flavor due to its large surface area.

UNREFINED SALT: Salt that is left in its natural state is unrefined. When it is mined or extracted it is only washed and strained. Unlike refined salt, all trace minerals are left intact.

Sea salt: As the name implies, sea salt is a broad reference to an edible salt that is drawn from the sea, the result of evaporated

continued on page 28

A Field Guide to Major Salt Types

seawater. Some people consider sea salt to be a healthier alternative to table salt because of the trace minerals it provides. There are many types of sea salt, including exotic finishing salts that are valued in the culinary world for their subtle differences in flavor, color, and texture.

Rock salt: Also known as halite, rock salt is the mineral form of sodium chloride. It is a mined salt and is most often inedible due to a lack of purification. Rock salt plays an important role in lowering the melting temperature of ice, whether on a sidewalk or when making homemade ice cream.

EPSOM SALT: Though it looks like salt and has salt in its name, Epsom salt is made up of the compound, magnesium sulfate, rather than sodium chloride. It was first distilled from the water of Epsom, England—hence the name—back in Shakespeare's day, and is most commonly used in bath salts and for medicinal purposes.

Do You Know ...

HOW VALUABLE SALT WAS TO THE ROMANS?

The word salary is derived from the Latin word *salarium*, the word for a soldier's pay in the army of ancient Rome. *Salarium* itself was derived from the word *sal*, or salt, because the pay included a large ration of salt. This in turn led to the expression, "worth your salt."

Kitchen Uses

FIRE FIGHTER

1 cup (300g) SALT

If a flare-up occurs while cooking, don't reach for the fire extinguisher; instead, quickly cover the bottom of the pan with a layer of salt (about 1 cup or 300g) and place a lid on top. Never try to tame the flames with water because that will only intensify the reaction.

SEE ALSO Stovetop Fire Fighter (baking soda) on page 168.

COPPER POT SCRUB

1/4 cup (60ml) VINEGAR
1 tablespoon (18g) SALT

To restore the shine to a copper pot, try this quick and easy duo. Soak a clean cloth in the vinegar and sprinkle with the salt. Rub over tarnished spots. You'll be amazed at how quickly the tarnish disappears. Rinse thoroughly with water and dry with a soft cloth.

SEE ALSO Copper Pot Scrub (lemons and salt) on page 76.

CAST-IRON CLEANER

$^{1}/_{2}$ cup (150g) SALT

$^{1}/_{4}$ cup (60 ml) vegetable oil

Cast-iron cookware can last a lifetime if cared for properly. To maintain a nonstick surface on cast-iron pans and to keep them in tip-top shape, always clean them with salt with this simple process: Rinse the skillet with hot water and scrape any residue with a wooden spoon or silicone spatula. Place the skillet over medium heat for 30 seconds; once dry, remove from the heat. Add the salt and vegetable oil and scrub with a thick wad of paper towels until clean. To finish, rinse again in hot water and dry over medium heat.

ENAMEL PAN CLEANER

2 tablespoons (36g) SALT

2 tablespoons (28g) BAKING SODA

Water

Some cast-iron cookware is coated in colored enamel, which not only makes it strikingly beautiful but also rustproof and nonreactive. The downside is that over time this type of cookware is prone to discolorations. To gently remove stains, mix equal parts salt and baking soda and add a small amount of water to make a paste. Apply the paste to the cookware and scrub with a nylon sponge. Rinse thoroughly with water and dry with a soft cloth.

WINE CHILLER

1 bucket ice

Water

½ cup (150g) SALT

Don't have that bottle of white wine chilled for your dinner guests? Cool it down quickly with the help of a little salt. Fill a bucket of ice with some water and add the salt. Place the wine bottle in the center and your wine will be perfectly chilled in as little as 10 minutes. That's because, just like when you're making homemade ice cream, salt lowers the freezing point of the ice, making the water surrounding the ice even colder than normal.

REFRIGERATOR CLEANER

¼ cup (75g) SALT

¼ cup (55g) BAKING SODA

Freshen up your refrigerator with the powerful combination of salt and baking soda. Mix the two together and sprinkle a small amount over any spills that have collected. Soak a sponge in some hot water and scrub gently. The baking soda and salt will work as a gentle abrasive that won't scratch refrigerator surfaces.

SEE ALSO Refrigerator Odor Eater (baking soda) on page 161.

CUTTING BOARD CLEANER

¹/₄ cup (75g) SALT

Wooden cutting boards are particularly sensitive to picking up stains and smells from the foods you cut on them. To clean them quickly and easily, wipe down with a damp sponge and sprinkle with about ¹/₄ cup (75g) salt. Let stand for a few minutes and then, working in small sections, use the sponge to rub the board until the stains are lifted. Rinse thoroughly with hot water and allow to air dry.

SEE ALSO What Else Can They Do? (lemons) on page 79; Berry and Beet Stain Remover (lemons and salt) on page 81; and What Else Can It Do? (vinegar) on page 118.

PERCOLATOR COFFEE POT CLEANER

¹/₄ cup (75g) SALT

Use salt for an effective, odor-free way to clean a percolator-style coffee pot. Fill the empty pot with water and place about ¹/₄ cup (75g) salt in the basket where the coffee normally goes. Plug in and let the salt gently soften the coffee buildup as it "brews." Drain and disassemble the pot. Rub the interior and all parts gently with a paper towel to remove any residue. Rinse thoroughly with water.

GLASS COFFEE POT CLEANER

½ cup (150g) SALT

Ice

LEMON slices

Water

Here's a restaurant trick for cleaning a glass coffee pot without get-ting your hands wet. Simply fill the pot with salt, ice, a few lemon slices, and just enough water to allow the ice to swirl around easily (make sure the pot is not hot before doing this or the quick change in temperature could crack the glass). Move the pot around gently in a circular motion and the ice and salt will lift coffee stains from the bottom of the pot. The acidity of the lemon juice will help cut through the oily residue left by the coffee. Rinse thoroughly.

SEE ALSO Coffee Pot Cleaner (vinegar) on page 118.

What Else Can It Do?

REMOVE LIPSTICK STAINS FROM GLASSWARE

Even after a trip through the dishwasher, lipstick stains can linger on glassware. Loosen stains significantly before washing by rub-bing them gently with a few spoonfuls of salt. The salt will cut right through the stain before it has a chance to stick.

Do You Know...

One of the most amazing things about salt is that it never loses its taste. In cooking, this constant flavor state makes for extremely predictable results—but that can backfire if you use too much salt. However, if you've accidentally oversalted a dish, there are some adjustments you can make:

• **Soups and stews:** Add half of a raw potato to the pot and cook for 10 to 15 minutes; it will absorb some of the excess salt. Remove the potato before serving.

• **Salads and dressings:** Compensate by adding more acidic ingredients, like lemon juice, vinegar, or tomatoes, to the mix.

• **Casseroles and side dishes:** Either double the recipe, omitting any more salt, or, if an acidic ingredient like tomato is present, try adding a pinch of sugar to counteract the saltiness.

What Else Can It Do?

SWEETEN FRUITS

Adding a pinch of salt to fresh fruits, such as melons and strawberries, will help bring out their sweetness.

SINK-FULL-OF-BUBBLES DISSOLVER

1 handful SALT

Once you've finished hand-washing a load of dishes, there's always the sudsy sink to contend with. Instead of wasting more water to rinse the suds down the drain, try sprinkling a handful of salt over all of the bubbles. They'll quickly dissolve, and the salt will actually help fight any grease that remains. Give the sink one last quick rinse and you're done.

CLOGGED DRAIN CLEARER

1/2 cup (150g) SALT
1 cup (220g) BAKING SODA
1 quart (1 liter) boiling water

When the kitchen sink is slow to drain, a sludgy buildup in the pipes may be the culprit. Let a trusted combination of salt and baking soda take the place of harsh chemicals that may be bad for the pipes. Pour the salt down the problem drain, followed by the baking soda. Let this mixture sit for several hours (overnight is best). Follow this up by pouring the boiling water down the drain to rinse it free of sediment.

SEE ALSO Drain Clearer (vinegar and baking soda) on page 120 and Drain Clearer (baking soda) on page 182.

OVEN SPILL ABSORBER

1 small handful SALT

What to do when the filling of a pie or casserole has bubbled over? Reach for the salt! A quick layer of salt on anything that's spilled in the oven will make for much easier cleanup once the oven cools down. Just make sure to put the salt down before the spill has a chance to bake on.

SEE ALSO Greasy Oven Buildup Prevention (vinegar) on page 122.

EGG SPILL ABSORBER

2 handfuls SALT

Raw eggs can make an especially aggravating mess to clean up because the whites have a tendency to slide away from the sponge or towel, spreading the original mess even farther. For easy cleaning, sprinkle the spill with a generous amount of salt. After 15 minutes, you'll find that the same mess doesn't move around as much when you wipe it up. Eggs that have been thrown at cars require an entirely different method for effective cleanup (see Egged Car Cleaner on page 150).

HARD CHEESE FRESHENER

1 tablespoon (18g) SALT

$\frac{1}{2}$ cup (125ml) warm water

Investing in a large block of hard cheese, like Parmesan, can be so much more economical than buying a number of small packages—but not if your cheese gets moldy before you can use it all. Here's a time-honored way to keep hard cheese fresher longer: Make a brine solution by dissolving the salt in the warm water. Soak a piece of cheese cloth in the mixture, wring it out until it is just damp, and use it to wrap your block of cheese. Place the cheese on a small tray in the back of your refrigerator. The salt in the damp cloth will help prevent mold formation.

"Of all the flavors one eats, salt is indispensable; wherever one goes in the world, one's mother is dearest." –CHINESE PROVERB

EGG FRESHNESS TESTER

2 tablespoons (36g) SALT

2 cups (500ml) water

It's hard to know how fresh an egg might be just by looking at its shell, but you can determine if it's fresh with this simple test: Dissolve the salt into the water and gently lower an egg into the water. If the egg floats, it's past its prime (because this shows that enough time has passed since the egg was laid for air to penetrate its semi-permeable shell). If it sinks, it's still fresh enough to use.

POULTRY AND PORK BRINE

1 cup (150g) kosher SALT

1/4 cup (50ml) sugar

1 gallon (4 liters) water

Soaking meat in a saltwater brine before cooking is an especially easy way to add flavor and tenderness to chicken, turkey, and pork. Plus, it locks in moisture, making cuts that are ordinarily vulnerable to dryness especially juicy and succulent. For the most basic of brines, dissolve the salt and sugar in the warm water. Refrigerate the poultry or pork in this mixture, totally submerged, for at least 4 hours and up to 24 hours. Rinse and pat dry before cooking or grilling. Feel free to add other flavorings, such as apple cider vinegar, onions, bay leaves, and peppercorns, to create your own unique specialty. TIP: Use a meat thermometer to check for doneness because brining has a tendency to make foods cook faster.

Do You Know...

In the United States, the Bonneville Salt Flats and the Great Salt Lake in the state of Utah are remnants of ancient Lake Bonneville, a body of water roughly the size of Lake Michigan, that existed about 15,000 years ago, during the last Ice Age. Every winter a shallow layer of standing water floods the surface of the salt flats and then slowly evaporates during the summer months, while winds smooth the surface into a vast, nearly perfect flat plain of 159 square miles (412 square km). The unique surface has been the site of numerous land speed records, including Gary Gabelich's 1970 record with his rocket car, Blue Flame, when he attained a spectacular 622.4 miles (1,001.7 km) per hour. The United States federal government over-sees the Bonneville Salt Flats and public use is tightly regulated. In fact, it is especially important to stay off the salt surface when it is covered by water. When wet, the salt surface is soft and cars can easily damage it. Furthermore, the salt water is highly corrosive and can "short out" the electrical system of a car.

"Salt is born of the purest of parents: the sun and the sea."
—PYTHAGORAS, GREEK SCHOLAR (580-500 BC)

Laundry Uses

YELLOWED LINENS BRIGHTENER

1 gallon (4 liters) water
$^1/_4$ cup (55g) BAKING SODA
1 tablespoon (18g) SALT

Whitening yellowed linens with a commercial bleach solution can seriously weaken the fabric fibers. Instead, try this gentle, easy method for brightening them: Bring the water to a gentle boil in a large pot. Add the baking soda and salt and stir until dissolved. Add your yellowed linens and boil gently for an hour, making sure the fabric is completely submerged. Drain and rinse the linens thoroughly in cool water. For best results, allow to air dry in direct sunlight. You can double or triple the formula for larger items.

SEE ALSO T-Shirt Whitener (lemons) on page 85.

"With a grain of salt."
–PLINY THE ELDER, ROMAN SCHOLAR (23-79 AD)

COLOR-FAST LAUNDRY

1 cup (250ml) VINEGAR

1/2 cup (150g) SALT

Color-safe detergent

If you love the color of your new cotton sheets or towels and don't want to see them eventually fade, a little vinegar and salt can help retain their brightness. Before putting your new items to use, fill your washing machine with cold water, add the vinegar, salt, and detergent, and wait a few minutes to let the salt dissolve. Wash your brights with similarly colored items in small loads and the colors will be set for future washings. Double the amount of vinegar and salt if colors are particularly vibrant.

What Else Can It Do?

CONTROL THE SUDS

Sure, it's a funny sight gag in the movies to see a washing machine overflowing with bubbles after too much soap has been added, but it's no laughing matter in real life. And with laundry products more concentrated than ever, it's an easy mistake to make. If you find yourself in such a situation and see the washer starting to spill over, add 1 cup (300g) salt to the load right away. Repeat if necessary.

What Else Can it Do?

STOP A STAIN FROM SPREADING ON A TABLECLOTH.

When you notice a small spill on your favorite tablecloth, no need to stop your dinner guests from having a good time. Apply the same technique as described in the Egg Spill Absorber (page 36) and discreetly cover the spill with salt to absorb the liquid and stop the stain from spreading. Proceed to the laundry room as soon as your guests are out the door.

SEE ALSO Red Wine Remedy for Carpets (baking soda) on page 181.

NYLONS MATCHMAKER

1 cup (300g) SALT
2 quarts (2 liters) water

Here's a time-honored trick for dealing with nylon stockings that are somewhat similar in color, but don't exactly match. Heat 2 quarts (2 liters) water in a large pot and add the salt. When the salt dissolves, add your mismatched hosiery to the pot and bring to a gentle boil for about 10 minutes. Remove pot from heat and cool to room temperature. Drain and rinse well to remove the salt. Line dry as usual, and you should find all of your stockings have been rendered the same color.

Personal Uses

SPLINTER REMOVER

¹/₂ cup (60g) Epsom SALT
2 quarts (2 liters) warm water

Before you start digging away at a painful splinter, try easing the way with Epsom salt. Mix the Epsom salt in the warm water and soak the afflicted finger or foot for 15 minutes. The salt will help soften the skin without the wrinkling normally associated with a warm-water soak, affording easier access to the splinter.

Do You Know...

HOW IMPORTANT SALT IS TO THE HUMAN BODY?

Salt is vital to health as it regulates the fluid balance throughout the body. Salt cravings may be caused by trace mineral deficiencies, as well as by a deficiency of sodium chloride itself. Unfortunately, with the abundance of salty, processed foods available, eating too much salt is all too common, and has been associated with an increase in a number of health problems, including high blood pressure.

SOOTHING FOOTBATH

$^{1}/_{2}$ cup (60g) Epsom SALT

$^{1}/_{2}$ cup (60g) BAKING SODA

1 gallon (4 liters) warm water

Looking for a great way to relax tired, aching feet? Try an old-fashioned footbath: Simply mix the Epsom salt and baking soda into a large basin of warm water. Slip your feet in and soak for as long as feels comfortable. Not only will your feet feel and smell great, but this effective combination will soften calluses, too.

SEE ALSO Muscle Ache Reliever (vinegar) on page 133.

MOISTURIZING BODY AND FACIAL SCRUB

$^{1}/_{2}$ cup (60g) sea SALT

$^{1}/_{3}$ cup (75ml) almond oil

5 drops lavender essential oil (optional)

Pamper yourself at home with an all-natural skin care formula. Pour sea salt into a small plastic container and cover with the almond oil and lavender essential oil, if desired. (If you mix it, the salt will quickly settle on the bottom, but if you pour the oil over the salt, it will slowly cover it on its own.) To use, gently massage a small amount onto wet skin, taking care to avoid eyes and other sensitive areas. Rinse thoroughly and pat dry. Store remaining scrub—tightly covered—away from light and heat.

SEE ALSO Gentle Facial Scrub (baking soda) on page 175.

What Else Can It Do?

SOFTEN SKIN WITH A POST-BATH SALT SCRUB

Another effective way to a healthy glow is to use salt to slough away dead skin cells. After bathing, stand in the tub and rub a few handfuls of salt into your skin. Rough patches on elbows and knees respond especially well to this treatment (again, avoid the sensitive areas). Rinse thoroughly with a lukewarm shower and pat dry.

HAIR REVITALIZER

3 tablespoons (24g) Epsom SALT

3 tablespoons (45ml) hair conditioner

Epsom salt can help transform your regular hair conditioner into a once-a-month deep conditioning treatment that restores body and shine, as well as curls if you have permed your hair. To use, mix equal amounts of conditioner and Epsom salt in a small bowl. Heat the mixture in the microwave for 20 seconds, stirring well to combine. To apply, wash hair as usual and then work the warm mixture through your hair from scalp to ends. Leave on for 20 minutes and then rinse thoroughly with warm water.

SEE ALSO Hair-Clarifying Rinse (lemons and salt) on page 92; Herbal Hair Rinse (vinegar) on page 135; and What Else Can It Do? (baking soda) on page 176.

EYE PUFFINESS REDUCER

$1/2$ teaspoon (3g) SALT

1 quart (1 liter) warm water

Eating salty foods can lead to puffy eyes, but, ironically, briefly soaking your eyes in a gentle saltwater solution can make them less puffy. Here's how to do it: Add the salt to the warm water and dip two cotton balls in the mixture. Lie down on your back, close your eyes, and place the cotton balls over your eyes for 10 minutes. Relax. When you get up, your eyes should appear less puffy.

What Else Can It Do?

FIGHT A FLAKY SCALP

Dandruff is caused by the normal shedding of skin cells on the scalp, but it can be terribly unsightly. Excessive brushing can make the problem seem even worse. A gentle way to easily loosen flakes before shampooing is to scrub your scalp with table salt. The salt crystals are much smaller than the bristles on a brush, resulting in a finer polish, and they'll dissolve quickly once you wet your hair (this treatment is more difficult with longer locks). Follow up by using a little less shampoo than normal because salt may intensify the lather.

SEE ALSO Dandruff and Oily Hair Fighter (lemons) on page 91 and Dandruff Fighter (baking soda) on page 177.

SORE THROAT SOOTHER

1 teaspoon (6g) SALT
1 cup (250ml) warm water

The next time a sore throat has you down, try easing the pain with this simple formula: Pour the salt into the warm water and stir until dissolved. Gargle a swig of this mixture for 5 to 10 seconds (try not to swallow) and repeat. Used 2 to 3 times a day, the salt will help reduce inflammation and speed healing.

SEE ALSO Sore Throat Soother (lemons) on page 89.

LEECH REMOVER

1 handful SALT

If you happen to emerge from a refreshing dip in a freshwater swimming area with a few leeches stuck to your skin, a sprinkle of salt is all you need to get them to release their grip. They release a powerful anesthetic when attaching themselves to your skin, so you're not likely to feel a lot of pain, but to prevent infection, you should follow up with regular first aid as you would an ordinary cut or scratch.

SINUS CONGESTION RELIEVER

$\frac{1}{2}$ teaspoon (3g) SALT

$\frac{1}{4}$ cup (60ml) warm water

When cold and flu season hits, dealing with a stuffy nose can become incredibly irritating, especially when regular blowing doesn't seem to clear things up. However, a gentle dose of saline a few times of day can help ease swelling and loosen mucus. Simply dissolve the salt in the warm water. Let the mixture cool to body temperature and then pour a small amount into the palm of your hand. Gently sniff the water into the clogged nostril while you press the other nostril closed with your finger. Repeat if necessary.

Do You Know...

WHAT A NETI POT IS?

A neti pot is a ceramic pot that looks like a miniature teapot and is used for irrigating nasal passages. It is especially helpful to those with chronic sinus problems. Used for centuries in India as part of the Ayurvedic yoga tradition, neti pots have come mainstream. Use the Sinus Congestion Reliever (above) and follow the manufacturer's instructions when using a neti pot.

Household Uses

WALL PATCHER

2 tablespoons (36g) SALT

2 tablespoons (16g) cornstarch

5 teaspoons water

No need to buy a whole container of patching compound to fill a few nail holes in a wall when you can whip up a batch of this simple home formula instead. Combine the salt, cornstarch, and water in a small bowl and stir to make a thick paste. Apply the paste to the nail holes with a putty knife and let dry. Rub lightly with sandpaper until smooth and then paint to match your wall.

"Wit is the salt of conversation, not the food." –WILLIAM HAZLITT, BRITISH ESSAYIST (1778-1830)

ARTIFICIAL FLOWER CLEANER

1 to 2 cups (300 to 600g) SALT

Some artificial flowers can look amazingly real, but a thin layer of dust on seemingly fresh petals is always a dead giveaway. Conventional dusting methods can make the problem even worse by driving dirt deeper into the flower. A great way to dust a small arrangement is to fill a paper bag with 1 to 2 cups (300 to 600g) salt; a plastic bag works well, too, if the flowers will fit. Place the arrangement inside and secure the top. Give the bag a few gentle shakes and the salt will brush the leaves and petals free of the dust.

What Else Can It Do?

KEEP ARTIFICIAL FLOWER ARRANGEMENTS IN PLACE

Here's a cheap way to keep your flower arrangements picture perfect. It's especially handy if it's hard to use the usual arrangement tools because the mouth of your vase is narrower than the base. Simply fill the container with salt before arranging your flowers and then pour a small amount of water over the salt once you're done arranging. Make sure to keep the arrangement undisturbed for the next few days. As the water evaporates, the salt will harden and keep the stems of your creation exactly where you want them to be.

THERMOS AND COOLER DEODORIZER

3 handfuls SALT

Salt has natural deodorizing properties that can work wonders on the stale smells that develop in closed containers like thermoses and coolers. Simply toss a few handfuls of salt into the container and let it sit for 24 hours. Rinse clean, and the smell should be noticeably absent. Better yet, get in the habit of putting a little salt in your containers before storing them. Just make sure to rinse well before using them again.

SEE ALSO Jar Deodorizer (vinegar) on page 121.

EASY AIR FRESHENER

1 cup (300g) SALT
1/4 cup (8g) dried lavender
3 to 5 drops essential oil (optional)

Mix the salt, lavender, and essential oil, if desired, in a decorative bowl or jar for an inexpensive air freshener. The scent will be strong for the first few days; as it dissipates, you can simply give the jar a quick stir to refresh. The fragrance will last for about a month, after which you can use the mix in your bath!

SEE ALSO Lemon Air Freshener (lemons and baking soda) on page 97.

BRASS CLEANER

2 tablespoons (36g) SALT

2 tablespoons (16g) flour

1 to 2 teaspoons (5 to 10ml) VINEGAR

In lieu of a toxic brass polisher, combine equal parts flour and salt in a small bowl and add just enough vinegar to make a paste. Use a soft cloth to apply the mixture to the brass item (wear rubber gloves, as this can be a dirty process if the item is heavily tarnished). Rinse thoroughly with water and dry with a clean, soft cloth. This formula can be easily doubled for larger items. TIP: Before attempting to clean a tarnished brass item, hold a magnet up to it to see if it really is brass. If the magnet sticks, the object is likely brass-plated and could be easily damaged by the abrasiveness of salt.

SEE ALSO Brass Cleaner (lemons and baking soda) on page 96.

RUST REMOVER

1 teaspoon (6g) SALT

$1/2$ LEMON

Here's a quick and easy way to remove rust spots from small chrome appliances, utensils, or bicycle handles. Simply sprinkle salt on the cut side of the lemon half and rub over the rust spots. Repeat if necessary. Rinse thoroughly with water and dry with a soft cloth.

STRAW BROOM LIFE-EXTENDER

1 cup (300g) SALT
1 gallon (4 liters) warm water

Here's a tried-and-true way to make a new straw broom last longer. Before first use, dissolve the salt in a bucket of warm water. Let the broom bristles soak for 20 minutes, then drain and let the broom air-dry. This process will help stiffen the bristles and make the broom much more durable.

NO-DRIP CANDLES

$\frac{1}{3}$ cup (100g) SALT
2 cups (500ml) warm water

No-drip candles are a great idea, but they're typically more expensive than their regular-burning counterparts. You can render tapered candles practically drip-proof with a simple overnight soak in saltwater. Mix the salt and warm water in a plastic container wide enough to hold the candles. Stir until the salt is dissolved and add the candles. Put a can or other weight on top of the candles to keep them submerged and soak overnight. Let the candles dry completely before using. TIP: For best results, make sure the burning candle is perfectly upright and in a draft-free place (both factors can make a candle burn unevenly, which may result in dripping).

MUDDY CARPET CLEANER

1 handful SALT

The next time someone tracks mud onto your carpet, resist the urge to blot it dry, as this could drive dirt deeper into the fibers. Instead, toss a handful of salt on the mess and walk away. The salt will absorb the liquid, making it easy to vacuum up when dry.

BLOODSTAIN REMOVER FOR CARPETS

1 handful SALT
½ cup (125ml) cold water
½ cup (125ml) hydrogen peroxide

Need to treat a bloodstain on carpet? Start by covering the area with salt to help absorb some of the blood. Next, mix equal parts cold water and hydrogen peroxide in a small bowl. Dampen a clean cloth with the mixture and use it to dab at the salt until a clump forms. Gently blot the stain and repeat until the stain disappears. Rinse by blotting with a clean cloth dampened with fresh water. NOTE: Do not use warm water as this can set the stain, and test on an inconspicuous area first.

Outdoor Uses

GARDEN SLUGS BE GONE

1 SALT shaker

Outside of the garden, slugs serve an important purpose in helping to recycle organic matter and build soil; however, that also means that you'll probably find them chewing up your favorite garden plants in no time. That's why it's a good idea to carry a small salt shaker in your pocket when you examine your garden plants, preferably in the early morning (slugs like to feed at night). Because their bodies are composed of so much water, a light sprinkling of salt will stop them in their tracks by dehydrating them.

SEE ALSO Garden Pest Fighter (salt) on page 56 and Aphids Be Gone (lemons) on page 105.

GARDEN PEST FIGHTER

1 cup (125g) flour
$^1/_2$ cup (150g) SALT

Cruciferous vegetables, like cabbage, broccoli, and cauliflower, are particularly vulnerable to a type of garden pest known as the cabbage worm. If you happen to see these insects munching away, stir together the flour and salt and put it in a shaker. Every morning or evening, when the plants are damp with dew, dust the leaves to keep pests at bay.

SEE ALSO Garden Slugs Be Gone (salt) on page 55 and Ahpids Be Gone (lemons) on page 105.

NO-FUSS FLOWER POT CLEANER

1 handful SALT

Scrub brushes can scratch and soap residues can cling to a clay pot's porous texture. Instead, reach for a handful of salt to work in their place. Simply place the salt on a clean damp rag and scrub away to let the gentle abrasive power of salt clean your empty garden pots between plantings. Rinse well with clean water.

What Else Can It Do?

KEEP ANTS AT BAY

Keeping an army of ants away from your home is much easier than trying to get rid of them once they've invaded. Fortunately, salt is a natural deterrent to ants, so you can use it safely along windowsills and doorways (or any threshold ants might be inclined to cross). Avoid the temptation to sprinkle salt directly in your yard because it will kill the grass.

SEE ALSO Ants, Roaches, and Fleas Be Gone (lemons) on page 102 and Anthill Antidote (lemons) on page 105.

Do You Know...

A CITY IN EUROPE THAT IS KNOWN FOR SALT?

While many spots throughout Europe are notable for their salt production, the Austrian city of Salzburg stands out from the crowd, as the name literally translates as "Salt City," coming from the Germanic root for the word salt, *salz*. The nearby salt mine of Dürrnberg played a pivotal role in providing near limitless riches to the area and now remains a celebrated tourist destination.

POISON IVY KILLER

1/2 cup (150g) SALT

2 quarts (2 liters) warm soapy water

Use this simple formula to kill any poison ivy plants you may have in your yard: Combine the salt and soapy water and stir to dissolve the salt. Pour the liquid into a large spray bottle. Spray the leaves and stems of the poison ivy plant (if the infested area is large, you can pour the mixture directly onto them). Once dead, be careful when you remove and discard the plants, as the powerful poison on the leaves may still be active, and never burn poison ivy as this can release toxic fumes into the air. NOTE: This formula will kill all kinds of plants, so make sure to avoid spraying any plants you want to keep.

SEE ALSO What Else Can It Do? (vinegar) on page 148 and What Else Can It Do? (baking soda) on page 189.

> "A man must eat a peck of salt with his friend, before he knows him."
>
> —MIGUEL DE CERVANTES, SPANISH WRITER (1547-1616)

What Else Can It Do?

STOP FLARE-UPS ON A BARBECUE GRILL

Dousing a barbecue flare-up with water is problematic for a couple of reasons. First, most flare-ups occur when the cooking meat releases its fat, and putting water on a grease fire is never advisable because it can make the flare-up even worse. Second, if you do happen to put enough water on the fire to tame the flames, the charcoal can become too wet to finish cooking whatever was in progress. However, you can control flames by tossing a handful of salt on a flare-up and closing the lid for a minute or two. So be sure to keep a jar of salt by your grill just in case.

SEE ALSO Barbecue Flare-Up Tamer (baking soda) on page 191.

PREVENT WICKER FURNITURE FROM YELLOWING

1/2 cup (150g) SALT
3 cups (750ml) warm water

White wicker furniture that's been finished with an oil-based paint has a tendency to yellow with age, especially if it is stored in the dark during the winter. To fight the effects of aging, give your wicker furniture an annual scrub down with a stiff brush and a saltwater solution. Allow the furniture to dry completely (preferably in the sun) before storing.

SEE ALSO Lawn Furniture Cleaner (baking soda) on page 186.

CAR WINDOW ANTI-ICER

¹/₄ cup (75g) SALT

1¹/₂ cups (350ml) warm water

To avoid the burdensome scraping and shoveling associated with winter snowstorms, try using the same salt strategy that works on roadways when another round of the white stuff is on its way. Dissolve the salt in the warm water, and wipe it on your car windows. The salt will lower the melting point of ice and snow when it hits the car, making the digging out process much easier. TIP: Take a trip through the car wash as soon as the weather improves; prolonged exposure to salt can harm the paint surface and leave your car more vulnerable to rust.

What Else Can It Do?

REMOVE ICE FROM SIDEWALKS

If a surprise snowstorm hits before you have had an opportunity to stock up on rock salt, you can always use table salt in a pinch. In fact, all types of salt work to melt ice by interfering with the water's ability to form crystals, and, thus, lowering the actual melting point of the ice. Rock salt is preferable because it is more economical for covering large surface areas than its fine-grained counterpart, but go ahead and use table salt if you're in a bind. TIP: Salt works best at temperatures near freezing; in sub-zero weather, it is noticeably less effective.

PINE TAR REMOVER

1 teaspoon (6g) SALT

1 squirt liquid hand soap

1 teaspoon (5ml) LEMON juice

Whether pruning hedges or hauling around a Christmas tree, it's impossible to not get your hands sticky with the distinctive scent of pine tar. Instead of using turpentine to clean up your hands, try this natural alternative first. Place 1 teaspoon (6g) salt into your hand, followed by a squirt of liquid hand soap and 1 teaspoon (5ml) lemon juice. Add enough water to form a lather, and gently scrub clean. Make sure to follow with hand lotion because the salt and lemon juice can be a bit abrasive.

SEE ALSO Yard Worker's Hand Wash (lemons and salt) on page 104.

Do You Know...

WHERE MOST SALT PRODUCTION OCCURS?

The United States and China combined annually produce forty percent of the world's salt supply.

Pet Uses

FLEAS BE GONE

1 handful SALT

If your pets have been afflicted with fleas, the problem most likely has spread to your carpets. To battle the bugs, sprinkle a handful of salt evenly over your carpet before you go to bed (make sure to keep pets out of the area as they may be tempted to lick it up). In the morning, vacuum thoroughly and repeat once a week for 6 weeks.

SEE ALSO Ants, Roaches, and Fleas Be Gone (lemons) on page 102 and Natural Flea Spray (lemons) on page 107.

PET BED CLEANER

$1/2$ cup (150g) SALT

To give the cleaning process a boost without adding a lot of artificial fragrances, try adding a $1/2$ cup (150g) salt to the laundry. For best results during warmer weather, make sure to air-dry the bedding in the sun.

SEE ALSO Pet Bed Deodorizer (baking soda) on page 193.

FISH TANK SCRUBBER

1 handful non-iodized SALT

Over time, mineral deposits and other buildup can make the inside of your fish tank look a little cloudy. And what's the point of having fish if you can't enjoy watching them? Because fish are especially sensitive to environmental pollutants, it's best to avoid harsh chemicals when cleaning their tanks. Instead, use a small handful of non-iodized salt to scrub the glass clean. Make sure to rinse the tank well before refilling it.

GOLDFISH SPA TREATMENT

1 quart (1 liter) distilled water
1 teaspoon (6g) SALT

Even goldfish need a little pick me up now and then! Although goldfish are freshwater creatures, giving them a little dose of salt can do them some good. Next time you're getting ready to change the water in your goldfish bowl, dissolve 1 teaspoon (6g) salt in the distilled water and let your gold fish swim around in this mixture for about 15 minutes while you change the water in the tank. Just make sure to return them to their tank promptly (and don't add the salt water to the freshwater by mistake).

Kids' Activities

MAKE A SALT PAINTING

4 to 5 small bowls SALT
1 to 2 drops food coloring, assorted colors
Glue
Paper

Emulate the amazing sand painting traditions of many indigenous cultures with this fun, easy, and inexpensive salt painting activity. To begin, pour an equal amount of salt into a few small bowls (just a few tablespoons or 54g in each should be enough). Add a drop of food coloring to each bowl and let your kids stir the salt with their fingers until the color is evenly distributed. Coat a piece of paper with a thin layer of white glue and help them sprinkle on the colored salts to create various, colorful patterns and shapes. When they're done, a damp sponge should be all you need for easy cleanup!

MAKE SALT CLAY DOUGH

1 cup (300g) SALT
1¹/₄ cups (300ml) warm water
1 tablespoon (15ml) vegetable oil
3 cups (375g) flour

Looking for a way to make a rainy day seem brighter for the little ones? Whip up a batch of this salty craft dough and they can spend an entire afternoon rolling, cutting, and shaping it into all kinds of memorable keepsakes. To begin, mix the salt with the warm water and stir to dissolve. Add the vegetable oil and flour and stir slowly to combine. Knead until the dough is elastic and pliable (add a little more flour, if necessary). Add a few drops of food coloring, if desired, to lend some color to the dough. If you like, let the kids' finished shapes air-dry for a couple of days and then have fun all over again painting them with watercolor or acrylic paints. A thin coating of clear nail polish or shellac will help preserve the finished masterpieces.

"Salt is the policeman of taste: it keeps the various flavors of a dish in order and restrains the stronger from tyrannizing over the weaker."
-MALCOLM DE CHAZAL, MAURITIAN WRITER (1902-1981)

CHAPTER 2

Lemons

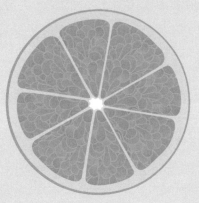

♦ ♦ ♦

The lemon *(Citrus limon)* is a yellow citrus fruit with a juicy acidic pulp, a sour flavor, a white inner pith, and an aromatic rind. Its main byproducts are citric acid (from the pulp), pectin (from the inner peel), and lemon oil (from the rind). Lemons are very high in ascorbic acid, or vitamin C. Lemon trees grow in subtropical climates; they have white flowers and thorny branches. There are several varieties of lemon, all quite similar, with the most common being the Eureka and the Lisbon. Other frequently grown types include the Villafranca, Verna, and Genoa. Meyer and Ponderosa lemons are actually hybrids.

A Brief History of Lemons

It's generally believed that lemons originated in Northern India, China, and Burma. Around 700 AD, lemons were introduced to the Middle East, and many scholars believe it wasn't until the last half of the first century AD that they came to the Mediterranean—when the Romans established a sea route to India.

Lemons began being cultivated in Europe, specifically in Genoa, Italy, in the mid-fifteenth century. Christopher Columbus, a native of Genoa, is credited with introducing the lemon to the Americas during his second voyage to establish a settlement on Haiti. Within twenty years of their introduction, lemons were growing there abundantly. The popularity of lemons naturally spread with subsequent colonization by the Spanish in the regions that are now California and Florida, and with the Portuguese colonization of Brazil. Today, lemons are grown around the world.

Lemons appear to have been used originally for their antiseptic properties and as ornamental trees. Although lemons are now considered one of the most important ingredients in many cuisines, they only appear to have gained widespread popularity within the last six hundred years.

The popular Meyer lemon (thought to be a cross between a Mandarin orange and a true lemon) was introduced to the United States in 1908 by Frank Meyer, an employee of the U.S. Department of Agriculture. The Meyer lemon is noticeably juicy and sweet; it is considerably less acidic than the traditional lemon variety. The Ponderosa is a cross between the true lemon and the citron; it produces extra-large fruit.

Do You Know...

WHAT TO LOOK FOR WHEN BUYING LEMONS?

Lemons should be firm and have a bright yellow color. Avoid soft, shriveled lemons or any with spots. The best lemons will be fine-textured and heavy for their size. Thin-skinned lemons tend to have more juice, while fruit that has a greenish cast is likely to be more acidic. One medium lemon has about 3 tablespoons (45ml) juice and 2 to 3 teaspoons (12 to 18g) grated peel, also known as zest.

How Lemons Work

Lemons contain five percent citric acid, just enough to impart a pleasant sourness that enhances the flavor of many foods. In cooking, lemon juice is generally interchangeable with vinegar and is frequently used with cooked fish. Owing to its citric acid, a little lemon juice sprinkled on freshly cut fruits, such as bananas, will prevent them from turning brown.

Around the house, lemons are used to neutralize alkaline stains, such as those caused by tea, coffee, perspiration, grass, and mustard. Lemons also have mild bleaching properties due to their high citric acid content, which make them ideal for using in the laundry and as hair rinses and beauty aides.

> **"It is probable that the lemon is the most valuable of all fruit for preserving health."**
> —MAUD GRIEVE (A MODERN HERBAL, 1931)

Kitchen Uses

ALL-PURPOSE LEMON CLEANER

2 to 3 cups (200 to 300g) leftover LEMON peel

2 cups (500ml) VINEGAR

What do you get when you combine the grease-fighting power of vinegar with the fresh scent of lemons? A perfect solution for all sorts of cleaning challenges! To make a batch, start by filling a large, wide-mouth jar with leftover lemon peels (other citrus fruits, such as oranges, limes, and grapefruits, work well, too). Cover the peels with the vinegar, place the lid back on the jar, and let sit in a dark, cool place for about 4 weeks (give the jar a good shake every week or so). Strain the mixture through a cheesecloth and discard the lemon peels. Use in place of regular vinegar for any of the cleaning formulas in this book.

SEE ALSO All-Purpose Baking Soda Cleaner (baking soda) on page 180.

MICROWAVE OVEN STEAM CLEANER

½ bowl water

½ LEMON

Even though microwave ovens aren't technically self-cleaning appliances, you can use the power of lemons to get a head start. Simply fill a microwave-safe bowl halfway with water. Add the lemon and heat on high for 2 minutes, until the water is boiling and giving off steam (power varies with microwave ovens, so adjust the timing accordingly). Keep the door closed for 5 minutes; when you open it, the microwave will be much easier to clean thanks to its citrus steam bath. Wipe down thoroughly with a nylon sponge.

CHEESE GRATER CLEANER

½ LEMON

Cheese graters are wonderful contraptions to have on hand if you love the taste of freshly grated cheese, but cleaning them afterwards is another matter. All of those small sharp edges tend to chew up your sponge or knuckles, or both! However, if you cut a lemon in half and rub the cut side on either side of the cheese grater after each use, the natural citric acid will cut right through the cheese that's left clinging to the metal. Rinse thoroughly with hot water and you're done!

FRUIT AND VEGETABLE WASH

1 cup (250ml) water

2 tablespoons (28g) BAKING SODA

1 tablespoon (15ml) LEMON juice

All produce needs a thorough washing before use, especially cucumbers and tomatoes, which typically have a waxy coating that can trap pesticides and other potentially harmful residues. Though there are plenty of produce washes on the market, you can skip the expense and whip up a homemade batch of your own. Fill a small spray bottle with the water, baking soda, and lemon juice (make sure to strain the juice if you're squeezing it fresh). Give the mixture a shake to ensure it's combined well, and then you're good to go. Spray on fruits and vegetables just before prepping or eating. To remove the bitter taste of baking soda, make sure to rinse the produce thoroughly with fresh water once you're done. Your homemade produce wash should last about 1 week.

SEE ALSO Fruit and Vegetable Bath (baking soda) on page 162.

"When fate hands you a lemon, make lemonade." –DALE CARNEGIE, AMERICAN WRITER (1888-1955)

What Else Can They Do?

MAKE A DELICIOUS SPRITZER

Simple syrup, which is basically sugar-sweetened water, makes it easy to whip up this delicious, fizzy alternative to lemonade. Combine ½ cup (100g) sugar and ½ cup (125ml) water in a small pan and cook over high heat until the sugar is dissolved to make enough simple syrup for several servings. Allow to cool to room temperature. Squeeze and strain the juice of a lemon into a large glass filled with ice. Pour an equivalent amount of syrup as to the lemon juice into the glass, fill with club soda, and enjoy!

DISHWASHING BOOST

2 tablespoons (30ml) LEMON juice

Plenty of dish soaps claim to cut the grease, but if you've happened to pick up a brand that doesn't seem to be living up to its promise, take matters into your own hands. Add about 2 tablespoons (30ml) lemon juice to your dishwater for an extra boost of grease-fighting power.

COFFEE AND TEA STAIN REMOVER

1 LEMON wedge
Boiling water

After years of use it's only natural that your favorite coffee mugs and teacups might start showing a few stains as a sign of their age. To get rid of these "permanent" stains, rinse out the cup, squirt a wedge of lemon into the bottom, and fill with boiling water (leave the lemon in the cup if you like). Soak overnight. In the morning, the stains should be visibly lightened.

COPPER POT SCRUB

$1/2$ LEMON
1 tablespoon (18g) SALT

To remove tarnish and restore the lustrous shine to copper cookware, use the cut side of a lemon as a handy scrubber. Simply sprinkle the cut side of the lemon half with the salt and rub over tarnished spots. The gentle acid from the lemon and abrasive power of salt will erase years of tarnish quickly and easily. Rinse thoroughly with water and dry with a soft cloth.

SEE ALSO Copper Pot Scrub (salt and vinegar) on page 29.

NONSTICK COOKWARE CLEANER

2 tablespoons (30ml) LEMON juice

1 teaspoon (5g) BAKING SODA

Sometimes a simple wash in hot, sudsy water doesn't remove all the oil that's left on a nonstick pan, especially if it's been used for frying. Eventually, thin layers of leftover oil will build up and become a gummy residue, causing food to stick when cooking. To ensure that your nonstick cookware stays truly clean, make a habit of regularly cleaning it with a bit of lemon and baking soda. Pour the lemon juice into the bottom of a just-cleaned pan and add the baking soda. Use a nylon-coated sponge to gently scrub. Rinse thoroughly with hot water.

What Else Can They Do?

REMOVE THE SMELL OF GARLIC FROM YOUR FINGERS

As any cook knows, the chore of chopping garlic can leave a strong scent on your fingers that regular soap and water just can't remove. By pouring a little lemon juice onto a stainless steel spoon, and then rubbing your fingers over the spoon for a minute or two, the smell should disappear. Just make sure you don't have any small cuts on your hands, as the lemon juice is sure to sting a bit.

SEE ALSO What Else Can They Do? (lemons) on page 79.

ALUMINUM POT AND PAN BRIGHTENER

3 tablespoons (45ml) LEMON juice
1 quart (1 liter) boiling water

Though it's a great material for conducting heat, aluminum is a relatively soft metal that's easily damaged by hard scrubbing. Plus, exposure to mineral deposits in food and water can leave the inside noticeably darkened or stained. One way to prevent this is to boil water with lemon juice inside the cookware. Follow this with a light rubbing using a soapy nylon-coated sponge. Rinse thoroughly.

PLASTIC FOOD STORAGE CLEANER

1 tablespoon (15ml) LEMON juice
1 tablespoon (14g) BAKING SODA

Tomato-based leftovers can stain plastic storage containers, especially if there's a bit of oil involved (think lasagna or chili). What's worse, these stains have a tendency to impart lingering food odors into other foods, interfering with their flavor. To give your plasticware a good, non-toxic scrubdown, start by mixing equal parts baking soda and lemon juice in the stained container. Scrub gently with a nylon-coated sponge to release the stains, but do not immediately rinse out the container; instead, fill with water and let sit overnight for maximum freshness.

GARBAGE DISPOSAL DEODORIZER

Rinds of 1 to 2 LEMONS

To rid your sink of bad smells and replace it with lemon freshness, juice 1 to 2 lemons and save the rinds to run through the garbage disposal with plenty of cold water. It's also a good way to dispose of older lemons that are past their prime and too soft to use in cooking; however, it's best to avoid lemons that have grown dark or moldy for these are not likely to smell fresh once cut open. TIP: What Else Can They Do? on page 75 has a delicious recipe that will utilize the left-over juice of the lemons.

What Else Can They Do?

DEODORIZE WOODEN KITCHEN UTENSILS

Wooden spoons and cutting boards are prone to picking up the scents of certain foods, like garlic and onions, and because of their porous nature it's best not to run them through the dishwasher. Instead, wash with a nylon sponge and then rub them down with the cut side of a lemon if strong smells remain. The scent of fresh lemon is a much better alternative to bleach.

SEE ALSO Cutting Board Cleaner (salt) on page 32; What Else Can They Do? (lemons) on page 77; Berry and Beet Stain Remover (lemons and salt) on page 81; and What Else Can It Do? (vinegar) on page 118.

What Else Can They Do?

MAKE FINGER FOODS LESS MESSY

Some of the best meals to share with family and friends are typically eaten with fingers (ribs and fried chicken, for example). But enjoying these foods during a more formal occasion can be tricky if you don't offer your guests a means of cleaning their hands between courses—after all, sometimes napkins alone won't cut it. Place a small bowl of water beside each guest's plate and float a few lemon slices in it. Your guests can easily rinse their fingers, and the lemon will effectively cut whatever oils may be on them from the food. (Just be prepared to enlighten anyone who thinks you're serving cold lemon soup!)

Do You Know...

WHAT OTHER FOODS TASTE LIKE LEMON?

Several other plants have a similar taste to lemons, including lemon myrtle, lemon thyme, lemon verbena, and lemongrass, but they don't contain lemon's characteristic citric acid. This makes them great substitutes in dishes with a tendency to curdle, such as cheesecake and ice cream.

The custom dates back to the Middle Ages when it was believed that if a person accidentally swallowed a fish bone, the lemon juice would dissolve it.

BERRY AND BEET STAIN REMOVER

1 tablespoon (15ml) LEMON juice

1 tablespoon (18g) SALT

Berries and beets contain natural pigments that enhance their health-promoting virtues, but the downside is that these same pigments can leave stains on your cutting board that are hard to get out. Fortunately, the duo of lemon juice and salt have a strong reputation for being the perfect cleaning team. The salt provides an abrasion while the lemon is a natural bleach and deodorizer. To use, combine equal parts of each, apply the mixture to the board, and scrub with a clean cloth. If you think the stain will be particularly stubborn, let the mixture sit for a few minutes before scrubbing. Rinse thoroughly once the stain has lifted.

SEE ALSO Cutting Board Cleaner (salt) on page 32; What Else Can They Do? (lemons) on page 79; and What Else Can It Do? (vinegar) on page 118.

Laundry Uses

ALL-NATURAL FABRIC SOFTENER

Juice of 4 LEMONS

2 cups (500ml) cold water

1 teaspoon (5g) BAKING SODA

For the freshest, cleanest linens and towels, turn to the remarkable power of lemons–not chemicals–to get the job done right. In a large plastic jar, combine the lemon juice with the cold water. Add the baking soda, cover, and shake well. Add the lemon mixture to the final rinse cycle and your sheets and towels will emerge feeling soft and smelling great.

SEE ALSO What Else Can It Do? (vinegar) on page 126 and Laundry Detergent Booster (baking soda) on page 169.

Do You Know...

HOW TALL A LEMON TREE GROWS?

An outdoor lemon tree can grow between 10 to 20 feet (3 to 6m) tall.

MOTHBALL ODOR REMOVER

1 cup (250ml) LEMON juice

Mothballs are made from naphthalene or paradichlorobenzene, chemicals that give off a noxious gas but that also prevent moths from chewing on woolen clothes. To rid your clothes of this strong, lingering scent, add 1 cup (250ml) lemon juice to the wash cycle when you launder them. Dry them outside, if possible, for best results.

UNDERARM STAIN REMOVER

1 tablespoon (15ml) LEMON juice
1 tablespoon (15ml) VINEGAR
1 cup (250ml) water

Perspiration stains often go unnoticed until they render clothing an unsightly yellow. Treat such stains with a soak in water with equal parts lemon juice and vinegar. Let soak for an hour or up to over-night, then launder as usual. NOTE: Never use a presoak product or formula on clothes made of silk, linen, or wool, as these types of fab-rics can develop water circles, bleed colors, or shrink. Also, if the stain remains after presoaking, never place the clothing in a dryer or iron it. This will cause the stain to become set by the heat, and, perhaps, make it impossible to get out.

SEE ALSO Perspiration Stain Remover (vinegar and baking soda) on page 124.

MILDEW STAIN REMOVER

2 tablespoons (30ml) LEMON juice
1 tablespoon (18g) SALT

If you happen to notice dark spots forming on your bathroom shower curtain, regular laundering tactics are unlikely to remove them, unless they're first treated with a lemon and salt scrub. Mix the lemon juice and salt and rub over the affected areas. Place in direct sunlight for the afternoon and then wash as usual. The stains should be gone without having to resort to harsh bleaches.

SEE ALSO Plastic Shower Curtain Cleaner (vinegar) on page 136 and What Else Can It Do? (baking soda) on page 181.

LEATHER SPOT CLEANER

1 tablespoon (15ml) LEMON juice
1 tablespoon (9g) cream of tartar

Try spot-cleaning stains on light-colored leather with a paste made from equal parts lemon juice and cream of tartar. Apply in an even layer and let sit for an hour or two before wiping off with a damp, clean cloth. NOTE: Make sure to test on an inconspicuous area first, as the lemon could lighten some types of dye.

SEE ALSO Leather Conditioner (vinegar) on page 142.

T-SHIRT WHITENER

¼ cup (60ml) LEMON juice

1 cup (250ml) water

Whether your once-white t-shirts have suffered a few accidental stains or are now just a tired shade of gray, a dose of lemon juice can dramatically restore their brightness. If you're dealing with specifically stained areas, apply straight lemon juice to the stains (no diluting required); however, if you're looking for overall brightening, mix the lemon juice and water in a small basin and soak the entire t-shirt in the mixture for up to an hour. Either way, allow the shirts to dry in the sun until desired whitening effects are achieved. Launder as usual to remove the juice.

SEE ALSO Yellowed Linens Brightener (salt and baking soda) on page 40.

What Else Can They Do?

WHITEN SNEAKERS

Have your white canvas sneakers seen better days? Instead of resorting to bleach to brighten them up, try spraying them with a light layer of lemon juice. Set them outside in direct sunlight and they should be noticeably whiter in a matter of hours.

BABY FORMULA STAIN REMOVER

¹/₄ cup (60ml) LEMON juice
¹/₄ cup (60ml) water

Baby formula is full of protein, so it can leave some nasty stains on your little one's clothing and blankets. For light-colored fabric, you can fight stains head-on without resorting to harsh bleaches by keeping a small spray bottle filled with lemon water in the fridge. Combine the lemon juice and water and apply to stained areas as soon as possible. Let air-dry in the sun and then wash as usual.

SEE ALSO Baby Clothes and Cloth Diaper Cleaner (baking soda and vinegar) on page 171.

PRE-TREATMENT FOR GRASS STAINS

1 tablespoon (14g) BAKING SODA
1 teaspoon (5ml) LEMON juice

How do moms know that spring has arrived? Their laundry baskets are suddenly overflowing with kids' jeans that are all sporting ground-in grass stains at the knees. Fortunately, there's a perfectly natural way to get rid of those stains. Sprinkle baking soda on the area, followed up with the lemon juice on top. As the mixture starts to fizz, rub the fabric to make sure it works its way through the stain. Launder as usual.

SEE ALSO Pre-Treatment for Kid Stains (baking soda and vinegar) on page 170.

LEMON-FRESH SHOE POLISHER

2 tablespoons (30ml) olive oil

1 tablespoon (15ml) LEMON juice

Here's a fresh-smelling polish that's also an eco-friendly way to maintain great looking shoes. Place the olive oil and lemon juice in a small lidded jar and shake vigorously until combined well (double the formula if you're doing several pairs of shoes). Dip a clean cloth into the mixture and apply evenly to dry, dark leather shoes that are in need of a bit of polish. Let the polish soak in for a few minutes, then buff with a fresh, clean cloth. Repeat monthly, or as necessary.

SEE ALSO Leather Shoe Cleaner (vinegar) on page 127 and Leather Shoe Conditioner (vinegar) on page 128.

Do You Know...

HOW MANY LEMONS GROW ON A LEMON TREE?

Mature lemon trees may produce between 1,000 and 2,000 fruits per year. Fortunately, this doesn't occur all at once! A lemon tree fruits all year long but most abundantly in late winter and spring.

Personal Uses

BAD BREATH FIGHTER

Juice of 1 LEMON
1/4 cup (60ml) water

Bad breath is a common problem that everyone worries about from time to time, especially after a meal laden with garlic and onions. If you don't have a toothbrush handy, a quick gargle with a lemon mouth rinse can help. Squeeze the juice of the lemon into the water (make sure to strain any seeds). Gargle this mixture for 15 to 20 seconds, taking care to swish it between your teeth and into the back of your throat. This solution is not harmful to swallow, but might be distasteful to some people.

What Else Can They Do?

RELIEVE A DRY MOUTH

Lemon juice can help kick the salivary glands into overdrive, making a wedge of lemon or even a little lemon candy the perfect antidote to dry mouth when a glass of water isn't handy.

SORE THROAT SOOTHER

1 teaspoon (5ml) honey

2 teaspoons (10ml) LEMON juice

Sore throats are often aggravated by a nasty buildup of phlegm, but a combination of lemon juice and honey is a great, natural way to gain some soothing relief. Place the honey in a small bowl and stir in the lemon juice. When completely combined, swallow the mixture slowly, allowing it to rest at the back of your throat as long as possible. The citric acid in the lemon juice will help to loosen congestion and the honey will provide a gentle coating. Both ingredients are credited with having antibacterial properties, too.

SEE ALSO Sore Throat Soother (salt) on page 47.

CONSTIPATION COMBATANT

1 teaspoon (5ml) LEMON juice

1 cup (250ml) warm water

1 teaspoon (5ml) honey

It's a condition people rarely talk about, yet one that affects many on a frequent basis. Before resorting to over-the-counter remedies for constipation, try this gentle cleansing formula: Add the lemon juice to a glass of warm water and then stir in the honey. Drink one glass of this lemon juice mixture to help move things along. Repeat every few hours, if necessary.

BLOND HIGHLIGHTS FOR HAIR

Juice of 2 or 3 LEMONS

For hundreds of years people have turned to the natural bleaching power of lemons to lighten their hair. Nowadays, it's a cheap and natural alternative to commercial products that contain ammonia. Comb the juice of 2 or 3 lemons through all your locks for an overall effect, or apply to select areas for a more sparing look. Sit outside in the sun or under a hair dryer until your hair is completely dry. Rinse out and style as usual. This simple approach will lighten all hair types, though darker hair may take on a brassy, versus blond, shade.

SWIMMER'S HAIR PREVENTION

1 12-ounce (355ml) bottle spring water
2 tablespoons (30ml) LEMON juice
2 tablespoons (28g) BAKING SODA

Long summer days spent cooling off in a chlorinated pool can impart a slight green tint to blond hair, a condition aptly known as "swimmer's hair." To rid your hair of chlorine after a dip in the pool, add the lemon juice to a bottle of spring water (take a few sips of water to leave room for the juice). Then stash in your swim bag along with a small plastic bag with the baking soda. After swimming, add the baking soda to the lemon water and shake well. Apply to your freshly shampooed hair, then rinse thoroughly and towel dry.

SEE ALSO Swimmer's Ear Prevention (vinegar) on page 133.

DANDRUFF AND OILY HAIR FIGHTER

$\frac{1}{2}$ cup (125ml) LEMON juice

1 cup (250ml) warm water

When a flaky scalp is caused by a buildup of dry skin, the gentle effects of lemon can be just the right antidote. To use, mix together the lemon juice and warm water. Shampoo and rinse as usual, then pour the lemon water over your freshly washed hair, rubbing gently into the scalp. Leave on for a few minutes, then rinse thoroughly. Repeat weekly as needed. If oily hair is the only problem, dilute the mixture with an extra cup of water.

SEE ALSO What Else Can It Do? (salt) on page 46 and Dandruff Fighter (baking soda) on page 177.

Do You Know...

WHERE LEMONS GROW?

With over thirteen million tons of lemons grown every year around the globe, India, Mexico, Argentina, Brazil, Spain, China, United States, Turkey, Iran, and Italy are the top-producing countries.

HAIR-CLARIFYING RINSE

1 gallon (4 liters) water

1 cup (250ml) LEMON juice

1 cup (300g) Epsom SALT

If you use a lot of hairspray and gel products, it's a good idea to use this clarifying rinse from time to time to get rid of the buildup that can weigh down your tresses. Fill an empty gallon-size milk jug with water, leaving some room at the top. Add the lemon juice and Epsom salt. Cover and let sit for 24 hours. Pour entire mixture slowly onto dry hair and let sit for 20 minutes. Shampoo as normal.

SEE ALSO Hair Revitalizer (salt) on page 45; Herbal Hair Rinse (vinegar) on page 135; and What Else Can It Do? (baking soda) on page 176.

AGE SPOT ANTIDOTE

$\frac{1}{2}$ teaspoon (2g) sugar

$\frac{1}{2}$ LEMON

Age spots are honestly nothing to worry about, but they can be unsightly for those afflicted. The good news is that they respond to the gentle bleaching abilities for which lemons are legendary. To use, sprinkle the sugar on the cut side of the lemon half and then gently rub over the affected area for a few minutes. Repeat weekly, as necessary, until the dark spots fade.

BLEMISH AND BLACKHEAD TREATMENT

1 teaspoon (5ml) LEMON juice

1 teaspoon (5ml) rose water

While there are a number of tried-and-true methods for treating black-heads and whiteheads, this all-natural formula will also combat oily skin, a condition that can contribute to blemishes. To use, mix equal parts rose water and lemon juice. Apply directly to skin with a cotton ball, concentrating on affected areas, and leave in place for 15 to 20 minutes. Rinse off completely and pat dry. Repeat 2 or 3 times a day until the blemishes have disappeared. TIP: Besides health food stores, rose water can often be found at Middle Eastern grocery stores.

What Else Can They Do?

PREVENT SCURVY

When deprived of vitamin C for months at a time, the human body will develop scurvy, a disease characterized by weakness, swollen joints, inflamed gums, and loose teeth. If left untreated, scurvy can lead to severe anemia. Scurvy was often the dreaded fate of sailors who lived at sea for months at a time without access to fresh fruits and vegetables. In 1795, British Royal Navy ships made it a practice to issue lime juice to their sailors based on the advice of the Scottish physician, James Lind. As they are an even better source of vitamin C, lemons and oranges eventually replaced limes as the preferred means for preventing scurvy.

CORN REMOVER

Warm water

1 LEMON wedge

Corns are essentially small, sometimes painful, patches of thickened skin that can crop up when you wear shoes that are too tight. To restore your feet, try this overnight strategy: Begin by soaking your feet in warm water for 10 to 15 minutes, then place a very small wedge of lemon over the corn. Secure the lemon with a bandage and leave it in place until morning. The mild acids from the lemon will gently soften the corn. This same strategy will work wonders for warts, too.

SEE ALSO Wart Remover (vinegar) on page 132.

CHEWING GUM REMOVER

1/2 LEMON

Nothing can be more frustrating than a cumbersome wad of gum in your hair and on your clothes. While a dab of peanut butter remains a popular and effective remedy, it can also cause some stains in its own right due to its high fat content. Instead, try rubbing the cut side of a lemon on the gum to help it loosen its grip (squeeze gently as you go to release the juice). Just be sure to rinse thoroughly to avoid the bleaching effects lemons can have.

NAIL BRIGHTENER

1 small bowl LEMON juice
¹/₄ cup (60ml) VINEGAR
¹/₄ cup (60ml) water

Tired of yellowing fingernails? Put the natural bleaching power of lemon juice to good use with this manicurists' trick! Soak your fingertips in a small bowl of lemon juice for 10 minutes (don't try this if you have any cuts or ragged cuticles). Next, pour out the lemon juice, rinse the bowl, and fill it with equal parts vinegar and water. Dip a soft nail brush in the vinegar solution and lightly buff your nails for a minute or two. Rinse thoroughly with warm water and follow up with your favorite hand moisturizer.

SEE ALSO Home Manicure or Pedicure Soak (baking soda) on page 176.

Household Uses

BRASS CLEANER

$\frac{1}{2}$ LEMON

1 tablespoon (14g) BAKING SODA

It's easy to fashion half of a lemon into an amazingly effective yet gentle all-natural scrubber for tarnished brass. To use, sprinkle the cut side of the lemon half with the baking soda and rub over tarnished brass objects. Rinse thoroughly with water and dry with a clean soft cloth.

SEE ALSO Brass Cleaner (salt and vinegar) on page 52.

Do You Know...

HOW TO GET THE MOST JUICE OUT OF A LEMON?

If the lemon has been refrigerated, warm it in the microwave for 15 seconds or in a pot of warm water for 15 minutes. If it is already at room temperature, try rolling it with your hand on the counter to release more juice.

LEMON-CINNAMON SIMMERING POTPOURRI

1 small pan water

1 cinnamon stick

1 LEMON, sliced

Here's an easy way to get that wonderful just-baked-something-delicious smell wafting through your home without turning on the oven! Make an aromatic simmering potpourri by filling a small pan with water and adding the cinnamon stick and lemon slices. Simmer on the stovetop, adding more water as necessary, and everyone will start hoping dessert is just around the corner.

LEMON AIR FRESHENER

2 tablespoons (28g) BAKING SODA

2 cups (500ml) hot water

1/2 cup (125ml) LEMON juice

Looking for an alternative to expensive room deodorizing sprays? Make your own! Dissolve the baking soda in the hot water, then add the lemon juice and transfer to a clean spray bottle. Use a few squirts now and then to fill your house with a just-cleaned scent (it's especially effective at clearing the air of dust after cleaning).

SEE ALSO Easy Air Freshener (salt) on page 51.

LEMON SACHET

1 cotton handkerchief

1 cup (220g) BAKING SODA

Zest of 1 LEMON

6-inch (15cm) piece of ribbon

Sachets are a nice way to keep spaces that are often dark and closed up, like drawers, storage containers, and closets, smelling fresh and sweet. If you don't care for sewing, you can easily and quickly fashion a sachet out of a plain cotton handkerchief and some ribbon. Lay the handkerchief on a flat work surface. Combine the baking soda and zest in a small bowl. Put the mixture in the middle of the handkerchief and gather the four corners of the handkerchief together so that the mixture assumes the shape of a ball. Twist the gathered corners and tie with the ribbon just above the ball. When the sachet loses its scent, use the baking soda mixture for a Drain Clearer on page 182.

SEE ALSO What Else Can It Do? (baking soda) on page 170.

Do You Know...

HOW TO ZEST A LEMON?

Zest, the colorful part of citrus peel, is valued for the strong flavor and scent from its aromatic oils. To remove the zest from a lemon, use either a microplane grater, a sharp paring knife, or a vegetable peeler. Remove only the colored portion of the peel (the white pith can be bitter) and then mince or leave in strips as directed.

HUMIDIFIER DEODORIZER

2 tablespoons (28g) BAKING SODA

3 tablespoons (45ml) LEMON juice, divided

A home humidifier can be a great tool for coping with the dryness associated with winter weather, but over time it can develop odors. If your humidifier is putting out more than just moisture, give all the washable parts a good scrub down with a mixture of the baking soda and 2 tablespoons (30ml) lemon juice. Rinse thoroughly and fill with the recommended amount of water, plus the remaining 1 tablespoon (15ml) lemon juice. The lemon will help transform the vapors into a pleasant lemony scent. Repeat every few weeks as necessary.

LEMON-FRESH TUB SCRUB

1/4 cup (55g) BAKING SODA

1/2 LEMON

Looking for an easier way to clean the tub? Sprinkle the baking soda on the bottom of the bathtub (use a little more if it's particularly dirty) and use the cut side of the lemon half as your scrubbing tool. Squeeze gently to release the juice as you scrub, and the combination of lemon and baking soda will cut through even the most stubborn bathtub rings. Rinse completely with water when you're done.

WINDOW AND MIRROR CLEANER

1 tablespoon (15ml) LEMON juice
2 cups (500ml) water
Newspaper

Here's a lemon-fresh way to clean your windows to a streak-free shine. Mix the lemon juice with the water in a large spray bottle (double the formula if the bottle is larger). Spray directly onto windows and mirrors and wipe clean with a crumpled wad of newspaper (use those old papers you've been meaning to recycle). Just remember to avoid cleaning windows while they are in direct sunlight, as the cleaning solution will dry too fast and lead to streaking.

SEE ALSO Window and Mirror Cleaner (vinegar) on page 138.

OLD-FASHIONED FURNITURE POLISH

$\frac{1}{2}$ cup (125ml) LEMON juice
1 cup (250ml) olive oil

For wooden furniture with an old-fashioned finish (i.e., an oiled instead of a shiny varnish), make your own furniture polish. Mix the lemon juice and olive oil in a glass jar with a lid. Cover and shake vigorously to combine, then apply to your furniture with a soft cotton cloth. Add a little shine by rubbing briskly, then allow to air dry. Refrigerate any unused portion for up to 3 weeks. TIP: The oil may appear cloudy when cold, but just let the polish sit at room temperature for a few minutes and it will return to normal.

DOORKNOB SANITIZER

½ cup (125ml) vodka

Peel of 1 LEMON

During cold and flu season, germs thrive on frequently touched surfaces, like doorknobs and toilet bowl handles. When someone in the house is under the weather, minimize the risk for everyone else by keeping this easy sanitizing spritzer on hand. Pour the vodka into a spray bottle and add the lemon peel. Store in a cool, dark place and the lemon will gently infuse the vodka with its clean, refreshing scent. Better yet, the alcohol content of the vodka will effectively kill germs when you spray a little bit in their direction.

What Else Can They Do?

NEUTRALIZE THE SMELL OF VINEGAR-BASED CLEANERS

Just as lemons contain citric acid, vinegars contain acetic acid, making both ideal ingredients in a variety of cleaning formulas. However, because of their essential oils, many people find the scent of lemons to be preferable to the more tangy scent of vinegar. If you want to use a vinegar cleaning formula (e.g., Sink and Countertop Disinfectant on page 137) but find the smell to be off-putting, squeeze a bit of lemon peel into the vinegar beforehand. Alternately, you can peel a handful of lemon zest strips and add them to a bottle of plain white vinegar. Store in a cool, dark place and the essential lemon oils will help tame the vinegar smell.

ANTS, ROACHES, AND FLEAS BE GONE

Juice of 4 LEMONS, or more for ants

2 quarts (2 liters) water

Just as vampires allegedly loathe garlic, many insects appear to despise the power of lemon juice, especially roaches, fleas, and ants. To get rid of roaches and fleas, make an effective floor-washing solution by adding the lemon juice (along with the rinds) to the water. Scrub the floor thoroughly and watch the critters scatter. Alternately, if ants are your problem, squirt fresh lemon juice around all their points of entry (doorways, windowsills, small holes, and cracks along the floorboards). For extra measure, chop the lemon peels into small pieces and leave them just outside your door.

SEE ALSO What Else Can It Do? (salt) on page 57; Fleas Be Gone (salt) on page 62; and Anthill Antidote (lemons) on page 105.

Do You Know...

WHY LEMON TREES ARE MORE SENSITIVE TO THE COLD THAN ORANGE TREES?

A lemon tree is less able to recover from a severe drop in temperature, such as a freeze, because it produces fruit all year round.

ATTRACTIVE LEMON CENTERPIECE

6 to 12 fresh LEMONS

1 bunch flowering herb, such as lavender or rosemary (optional)

Why focus on expensive, fresh floral arrangements when you can create centerpiece magic with a bag of lemons? Fill a decorative vase or large bowl with lemons and fill in the spaces along the bottom with sprigs of fresh herbs. You'll have a colorful, fresh-smelling arrangement your guests are sure to admire and remember—and you'll have a lot of uses for those lemons later!

What Else Can They Do?

ADD A CITRUS SCENT TO KINDLING

If you happen to have a few lemon peels left over, don't throw them out! Cut them into a few big pieces and put them in a safe place to air-dry (the bottom of a cabinet works well if you line it with paper towels first). The next time you are ready to build a fire, add some of the lemon peels to the kindling mix. They'll add a wonderful citrus scent to the fire!

Outdoor Uses

YARD WORKER'S HAND WASH

1 teaspoon (6g) SALT

$1/2$ LEMON

An afternoon of gardening or yard work may leave you with sticky plant residues on your hands that can be difficult to scrub off with the usual soap and water routine. Fortunately, you don't have to resort to harsh cleansers. Just put the salt in the palm of one hand, squeeze the lemon half over it, and gently rub your hands together (don't try this if you have any cuts or ragged cuticles). The salt is a natural abrasive, and the lemon juice is effective at cutting through the tar-like properties of many plant compounds. Rinse thoroughly and pat dry. Follow up with your favorite hand moisturizer if they feel slightly tender.

SEE ALSO Pine Tar Remover (salt and lemons) on page 61.

APHIDS BE GONE

1 tablespoon (15ml) LEMON juice

2 cups (500ml) water

1 tablespoon (15ml) baby shampoo

Your garden could be in big trouble if your plants are showing signs of an aphid infestation. To keep things under control, mix up a batch of aphid spray by combining equal parts lemon juice and baby shampoo in 2 cups (500ml) water. Spray the mixture on the affected leaves and your plants will be much less inviting to the aphids.

SEE ALSO Garden Slugs Be Gone (salt) on page 55 and Garden Pest Fighter (salt) on page 56.

ANTHILL ANTIDOTE

1 handful LEMON peel

1 cup (250ml) warm water

If you have a serious problem with ants, the power of lemon will persuade them to move to another location. Throw a handful of chopped lemon peel into the food processor, and while the machine is running, slowly add the water to make a lemony liquid. Carefully pour the solution over and into any problem anthills.

SEE ALSO What Else Can It Do? (salt) on page 57 and Ants, Roaches, and Fleas Be Gone (lemons) on page 102.

MOSQUITO REPELLENT

Peel of 1 LEMON

Few things can disrupt the pleasantries of summer like a slew of itchy and unwelcome mosquito bites. If you prefer to try a natural way to keep the mosquitoes at bay, cut the peel off a lemon and then rub the essential oil from the outside (yellow side) on your skin. TIP: If you fold a piece of peel between your thumb and forefingers (yellow side out) and give a gentle squeeze, the oils will be released. You'll actually see a squirt and smell a strong lemon scent when that happens.

SEE ALSO What Else Can It Do? (vinegar) on page 149 and Bug Bite Remedy (baking soda) on page 175.

What Else Can They Do?

SATISFY A CRAVING FOR SALT

Because of their tartness, lemons are a flavorful substitute for salt in cooking, especially if you are on a low-sodium diet. And if you are on a low-fat diet, substitute lemon juice for butter.

Pet Uses

NATURAL FLEA SPRAY

1 LEMON

2 cups (500ml) water

To prepare a homemade flea spray, cut the lemon into 4 to 6 pieces and place in a large glass bowl. Bring the water to a boil and pour over the lemon pieces, then let the mixture sit overnight. In the morning, strain the liquid and transfer to a spray bottle. Use to spray your dog or cat generously from head to tail, especially behind the ears (be careful to avoid the eyes).

SEE ALSO Fleas Be Gone (salt) on page 62 and Ants, Roaches, and Fleas Be Gone (lemons) on page 102.

Do You Know...

WHY LOUIS XIV GIFTED HIS MOST FAVORITE WOMEN IN HIS COURT WITH LEMONS?

At the time, it was fashionable to redden one's lips with lemons.

CAT DETERRENT

1 handful LEMON peel

1 handful coffee grounds

If the neighborhood cat has mistaken your garden for a litter box, lemon peels will make the area a lot less inviting. Use a food processor to chop up a handful of peels and then mix them with your leftover coffee grounds. Scatter the mixture along the border of your garden and work it into the soil with a hand trowel. Cats simply hate the smell of citrus and should steer clear. If your indoor plants are falling victim to your cats, this method should work for them as well.

SEE ALSO Cat Behavior Moderator (vinegar) on page 151.

"We are living in a world today where lemonade is made from artificial flavors and furniture polish is made from real lemons." -ALFRED E. NEUMAN, ICON OF *MAD MAGAZINE*

Kids' Activities

MAKE INVISIBLE INK

1 piece of paper
1 cotton swab
1 small bowl LEMON juice

Write a message on a piece of paper with a cotton swab dipped in lemon juice as invisible ink. After the "ink" is dry, hold the paper near a hot light bulb (not too close!). The writing will turn brown and you'll be able to read the secret message.

SHINE PENNIES

1 small bowl LEMON juice
Pennies or copper coins

The acidic property of lemon juice is great way for transforming dingy pennies into shiny coins. Just soak your pennies in a bowl of lemon juice for 15 to 20 minutes. Remove and let dry on a paper towel, and your pennies will look brand new!

CHAPTER 3

Vinegar

Vinegar is made when bacteria called acetobacters convert a fermented liquid into a weak solution of acetic acid. To understand how this transformation works, it's helpful to remember that vinegar is typically the byproduct of several natural processes that involve bacteria. Firstly, bacteria convert a food containing sugars into alcohol through a process called fermentation (that's how grapes become wine; grains become beer), and then another group of bacteria step in and convert the alcohol portion of the liquid into acetic acid. The presence of acetic acid is what essentially defines vinegar.

A Brief History of Vinegar

Vinegar wasn't really invented as much as it was discovered, 10,000 years ago, most likely when a flask of wine was left open and, exposed to the air, fermented further, causing the alcohol to turn to acetic acid. The name itself derives from the French words *vin* and *aigre*, meaning "sour wine."

Vinegar has been in common use for thousands of years. Traces of it have been found in ancient Egyptian vessels. Mentions of vinegar appear in numerous ancient writings from Chinese herbals to the works of Hippocrates. It was used for pickling and preserving, to treat infections and wounds, as an antiseptic and deodorizer, as a condiment, and even as a beverage.

Hannibal used hot vinegar to crumble large boulders as he moved his army through the Alps; thieves doused themselves in vinegar to fight off germs before robbing the dead during the Black Plague of the fourteenth century; people held vinegar-soaked sponges to their noses to diminish the smell of sewage in the streets in the seventeenth century; and doctors treated wounds with it on the battlefields of World War I.

Not surprisingly, given the demand and the inconsistency of home-made efforts, vinegar-making became a commercial undertaking as early as 2000 BC. By the late 1300s, in France, master vinegar makers had developed the Orleans method, whereby they could make continuous batches of vinegar by adding fresh cider or wine to oak barrels containing the remnants of the previous batch ("the mother of vinegar.") Flavored vinegars soon followed, as makers infused herbs, fruits, and spices into their concoctions. In 1869, H. J. Heinz was the first to mass produce and distribute vinegar.

"Marriage from love, like vinegar from wine—
A sad, sour, sober beverage—by time
Is sharpen'd from its high celestial flavour
Down to a very homely household savour."
–LORD BYRON, BRITISH POET (1788-1824)

How Vinegar Works

Just as the citric acid in lemon is responsible for the fruit's characteristic sourness and stain-fighting ability, vinegar owes its many beneficial properties to an acid—specifically acetic acid. Commercial vinegar typically ranges in concentration from four to seven percent acetic acid, and it is precisely within this very low-strength window that vinegar is able to be both palatable and potent. Stronger concentrations of vinegar are not considered safe for home use. Few vinegars are meant to be consumed "straight"—they tend to temporarily shut down the taste buds—but are balanced by combining with other foods.

In cooking, vinegar works as a preservative when making pickles, mustards, and other condiments (remember, bacteria can't grow in vinegar's acidic environment). Vinegar is also an essential ingredient in meat marinades, as the acid gently breaks down proteins and tenderizes the meat before cooking. In salad dressings, vinegar is a natural complement to oil.

Around the house vinegar works wonders because it can kill germs and also neutralize acidic and protein-based stains—like mustard, grass, and blood—by diluting them and not allowing them to take hold in a fabric.

Making vinegar at home is possible, but generally not advisable since so many factors influence the quality of the final product; also, commercial versions are quite inexpensive and readily available. All types of vinegar are best stored airtight in a cool, dark place.

The Many Flavors of Vinegar

In cooking, vinegar lends a general flavor of sourness to a dish, as well as a hint of whatever ingredients it originated from (a wine flavor in wine vinegar, or an apple note in a cider vinegar, for example). Many Asian vinegars are made from rice and present a very light, sweet flavor. And because vinegar absorbs other flavorings easily, infusing it with their flavors can create different types of vinegar; raspberry, tarragon, basil, and chili are some of the most popular flavored vinegars. NOTE: For general use and for the purposes of this book, use distilled white vinegar unless a formula directs otherwise; it won't stain and is most affordable.

Kitchen Uses

DISHWASHER WASH

5 or 6 VINEGAR ice cubes

This might sound like an odd household chore—after all, wouldn't you expect the inside of a dishwasher to be clean after washing all those dishes? But think about the little bits of food that can accumulate, and the fact that it's a dark container that is often closed and damp inside. To give your dishwasher an easy interior cleanup, freeze some vinegar in an ice cube tray. Place 5 or 6 cubes in the bottom of your dishwasher and run it on its highest heat setting without any dishes inside.

What Else Can It Do?

TURN MILK INTO BUTTERMILK

If you have a recipe that calls for buttermilk, don't bother running out to the store if you don't have any on hand. Just stir 1 tablespoon (15ml) vinegar into 1 cup (250ml) milk. It will thicken the milk instantly and lend the characteristic sourness of buttermilk to any dish.

COFFEE POT CLEANER

2 cups (500ml) VINEGAR

1 quart (1 liter) water

To make sure your drip-filter coffee pot produces a perfect brew every time, it's important to give it a deep cleaning about once a month to get rid of the mineral deposits and natural coffee oils that can build up. Pour the vinegar and water into the water reservoir and turn the coffee maker on (adjust this amount depending on the size of your machine; just keep the ratio 1 part vinegar to 2 parts water). When the cycle is complete, turn off the coffee maker and let cool for 15 to 20 minutes. Pour the vinegar mixture down the drain, fill the water reservoir with fresh water only, and run the coffee maker again.

SEE ALSO Glass Coffee Pot Cleaner (salt and lemons) on page 33.

What Else Can It Do?

DISINFECT WOODEN CUTTING BOARDS

To disinfect and clean your wooden cutting boards or butcher-block countertop, spray with full-strength vinegar after each use and wipe clean. The acetic acid in the vinegar is a natural disinfectant that is effective against the bacteria that can cause food-borne illnesses.

See also Cutting Board Cleaner (salt) on page 32; What Else Can They Do? (lemons) on page 79; and Berry and Beet Stain Remover (lemons and salt) on page 81.

TEAKETTLE CLEANER

½ cup (125ml) VINEGAR

2 cups (500ml) water

With repeated use, teakettles tend to develop a buildup of lime scale that's essentially the accumulation of minerals from hard water. Aside from drifting free now and then into your teacup, lime scale can also slow down the efficiency of the pot in reaching a boil. Fortunately, a quick vinegar steam is all it takes to significantly eliminate much of the problem. Boil the vinegar and water in the teapot for 10 to 15 minutes, remove from heat, and let cool. Drain and refill with fresh water and boil again for 10 minutes to remove all traces of vinegar. Repeat every few months (or more frequently, if you're an avid tea drinker) to keep your pot in tip-top shape.

WINE GLASS SPARKLER

1 quart (1 liter) VINEGAR

If your wine glasses have taken on a cloudy hue, you can easily revive them with a soak in straight vinegar. Arrange glasses in a small dishpan and add enough vinegar to the pan to keep them covered (use more or less depending on the number of glasses). Allow to soak for 1 to 2 hours. Rinse carefully, hand wash in warm water as usual, and dry with a lint-free cloth. TIP: To prevent new glasses from fogging up in the first place, add a little vinegar to the dishwater before washing them.

CRYSTAL VASE CLEANER

1 cup (250ml) VINEGAR

2 to 3 drops dish soap

A delicate crystal vase can be tricky to clean, especially if the mouth of the vase is relatively small. Whatever you do, don't put it in the dishwasher! Crystal is extremely vulnerable to hot water and can easily crack. Instead, fill the vase with the vinegar and add the dish soap. Let sit for an hour, then rinse thoroughly. The acetic acid in vinegar will gently clean your crystal and leave it sparkling like new.

DRAIN CLEARER

$1/2$ cup (110g) BAKING SODA

$1/2$ cup (125ml) VINEGAR

2 quarts (2 liters) water

A slow moving drain can be a harbinger of bigger problems unless it's dealt with swiftly. But before you pour harsh chemicals down the drain, give this easy, natural formula a try: Start by pouring the baking soda into the problem drain, followed by the vinegar and let sit for 15 minutes. Meanwhile, bring the water to a boil. After 15 minutes, pour the boiling water down the drain. Repeat monthly to keep your drain clean and smelling fresh.

SEE ALSO Clogged Drain Clearer (salt and baking soda) on page 35 and Drain Clearer (baking soda) on page 182.

JAR DEODORIZER

2 to 3 tablespoons (30 to 45ml) VINEGAR

Reusing glass jars with screw-on caps is economical and earth-friendly. They can come in handy for homemade gifts, craft projects, and storing other foods. What you don't want to reuse, however, is the lingering scent of the original contents. To get rid of unwanted smells, pour 2 to 3 tablespoons (30 to 45ml) vinegar into the bottom of the empty jar. Place the lid on top and swish the vinegar around. Let it sit for an hour, then wash as usual. If the scent is especially strong, let the vinegar sit overnight.

SEE ALSO Thermos and Cooler Deodorizer (salt) on page 51.

ALL-PURPOSE METAL COOKWARE SCRUB

1 tablespoon (8g) flour
1 tablespoon (18g) SALT
1 tablespoon (15ml) VINEGAR

Here's a simple-to-remember formula for an amazingly effective scouring scrub: Measure equal amounts of flour, salt, and vinegar and mix into a paste. You'll have an all-purpose cleaner that's safe for cleaning all of your metal cookware. Rinse off with warm water and dry thoroughly with a soft kitchen towel.

SEE ALSO Cleaner for Most Pots and Pans (baking soda) on page 164.

GREASY OVEN BUILDUP PREVENTION

½ cup (125ml) VINEGAR
½ cup (125ml) distilled water

After going through the effort to get your oven sparkling clean, protect your hard work and avoid the inevitable greasy buildup a little longer by coating the inside of your oven with equal parts vinegar and distilled water (use distilled water to avoid coating your oven with the minerals found in tap water). Apply with a clean cloth and let air-dry before using your oven.

SEE ALSO Oven Spill Absorber (salt) on page 36.

Do You Know...

WHAT IS THE MOST EXPENSIVE TYPE OF VINEGAR?

A bottle of true balsamic vinegar from Modena, Italy, can sell for several hundred dollars! Two main factors influence the price: its age (minimum 12 years up to 100 years—the longer it ages the more its quality improves, like a fine wine), and the very limited supply that is released every year. Most interestingly, even though it's considered a wine vinegar, it's actually made from the pressings of sweet white grapes; they're boiled down to a dark syrup and then barrel-aged. Fortunately, its strong flavor allows a small amount to go a long way.

FRUIT FLY TRAP

2 tablespoons (30ml) apple cider VINEGAR
1 small piece overripe fruit
Plastic wrap

If you happen to have a fruit fly infestation thanks to an overripe piece of fruit in the house, here's a surefire formula to get rid of those little gnat-like creatures. Make a trap by placing a small piece of over-ripe fruit in a bowl and covering with apple cider vinegar. Cover the bowl with a piece of plastic wrap and poke a dozen small holes in the plastic. The flies, attracted by the irresistible contents inside, will enter through the holes, but won't be able to exit.

"If you pour oil and vinegar into the same vessel, you would call them not friends but opponents."

—AESCHYLUS, GREEK PLAYWRIGHT (525-456 BC)

Laundry Uses

PERSPIRATION STAIN REMOVER

3 gallons (11 liters) water
1/4 cup (60ml) VINEGAR
1 tablespoon (14g) BAKING SODA

For serious perspiration stains, try soaking the item overnight in a tub filled with the water and vinegar (don't soak silk, acetate, linen, or colored cotton in vinegar, as these fabrics may be damaged). In the morning, launder in the washing machine as usual. For spot treatments, mix 1 teaspoon (15ml) vinegar with 1 tablespoon (14g) baking soda and rub into the stain just before washing. NOTE: Whatever you do, make sure the stain is properly removed before you put the clothing in the dryer. Heat will do a great job of setting the stain and actually making it appear darker and more pronounced.

SEE ALSO Underarm Stain Remover (lemons and vinegar) on page 83.

EASY IRON CLEANER

½ cup (125ml) VINEGAR

½ cup (125ml) water

1 tablespoon (14g) BAKING SODA (optional)

A dirty iron can spell disaster when burned-on buildup of starch and hard water deposits leave a stain on freshly laundered fabric. To keep your iron in prime condition, combine the water and vinegar, along with the baking soda if the iron is especially dirty. Wet an old but clean towel with the mixture, and then iron the towel until the plate of the iron is clean. Most manufacturers advise against putting anything except distilled water in an iron's reservoir, so don't use this formula from the inside out!

STARCH-FREE CREASER

⅓ cup (75ml) VINEGAR

⅔ cup (150ml) water

Parchment paper

Who says you need starch to put a permanent crease in your freshly laundered clothes? Mix 1 part vinegar with 2 parts water and transfer to a spray bottle. Squirt a little mixture on the section where you want a crease, cover with a piece of parchment paper, and iron as usual. Of course, if you prefer to remove a crease, rather than make one, use the same process but flatten the existing crease. This works especially well on twill pants and denim jeans.

FORGOTTEN LOAD RESCUER

2 cups (500ml) VINEGAR

Have you ever started a load of laundry only to realize later you have forgotten to transfer it to the dryer? If this happens during the dog days of summer, nature has a way of sending its own unpleasant reminder in the form of a distinct, moldy smell. Unfortunately, simply washing the clothes all over again may not be enough to get rid of the odor. Instead, start the washer again with hot water (no soap) and 2 cups (500ml) vinegar. After the first cycle, restart your machine with the usual amount of soap added.

What Else Can It Do?

GET YOUR LAUNDRY CLEANER AND SOFTER

Add 1 cup (250ml) vinegar to the last rinse of your washing machine cycle. The acid in vinegar is too mild to harm fabrics, yet strong enough to dissolve the soap residues that can make clothes feel stiff after drying. Besides removing soap, vinegar breaks down uric acid (so it's especially good for baby clothes), prevents yellowing, acts as a fabric softener and static cling reducer, and attacks mold and mildew.

SEE ALSO All-Natural Fabric Softener (lemons and baking soda) on page 82 and Laundry Detergent Booster (baking soda) on page 169.

COTTON AND DOWN PILLOW CLEANER

¹/₄ cup (60ml) VINEGAR

If you're troubled by allergies, don't overlook the importance of pillow cleaning. At least every 6 months you should wash pillows according to manufacturer's instructions to get rid of the sloughed skin, mold and mildew, fungus, and dust mite matter that can accumulate. When machine washing, add ¹/₄ cup (60ml) vinegar to the final rinse in lieu of fabric softener. It's the best way to make sure all the soap is removed. If you're washing a down pillow, add a tennis ball to the wash to keep the down from bunching up during the spin cycle, and spin dry twice, if possible, to ensure that most water has been removed.

LEATHER SHOE CLEANER

¹/₂ cup (125ml) water
¹/₂ cup (125ml) VINEGAR

In winter, a quick walk through a slushy parking lot is likely to leave quite a bit of salt on your shoes and what may be good for melting ice can actually damage leather, if you allow the salt stains to build up over the season. To make an effective shoe cleaning solution, mix equal parts water and vinegar. Dip a clean cloth into the solution and dab it sparingly over the salt-streaked parts of your shoes. Repeat, as necessary, until all the salt is removed, and then let them air dry.

SEE ALSO Lemon-Fresh Shoe Polisher (lemons) on page 87 and Leather Shoe Conditioner (vinegar) on page 128.

LEATHER SHOE CONDITIONER

1 tablespoon (15ml) VINEGAR

1 tablespoon (15ml) linseed oil

For regular cleaning and conditioning, an equal mix of vinegar and linseed oil can help keep leather shoes soft and supple. Mix the two together in a small glass container. Apply to shoes with a soft cloth and buff lightly. NOTE: Rags soaked in linseed oil are considered a fire hazard, so immediately soak them in water after use and never put them in the dryer.

SEE ALSO Lemon-Fresh Shoe Polisher (lemons) on page 87 and Leather Shoe Cleaner (vinegar) on page 127.

What Else Can It Do?

CLEAN SUEDE SHOES

Generally speaking, it's best to have all suede professionally cleaned, but in the case of shoe stains, vinegar may be worth a shot. Start by restoring the nap of the suede by brushing with an old bath towel. Rub a pencil eraser on any dark spots to lift dry stains, and blot wet stains with paper towels. Soak a corner of the bath towel with a few tablespoons (45ml) of vinegar and apply sparingly to remaining spots. Allow to air dry before wearing again. It's possible that the vinegar scent may linger, but it will eventually fade over time.

CIGARETTE ODOR REMOVER

1 quart (1 liter) VINEGAR

The none-too-pleasant scent of cigarette smoke has a shocking ability to cling to clothing, and it can even spread to other garments hanging nearby. But what to do when a conventional washing or trip to the dry cleaner isn't an option? Turn to vinegar, of course! To start, fill your bathtub with hot water and add 1 quart (1 liter) vinegar. Hang the affected clothes just above the tub so that as the vinegar-scented vapors rise, they'll eradicate the smoke smell. Keep the door closed for maximum effect and leave the clothes to hang overnight. In the morning, the clothes should smell noticeably fresher.

SHRUNKEN WOOL SWEATER RESTORER

1 quart (1 liter) VINEGAR

2 quarts (2 liters) water

Everyone has had the unfortunate experience of accidentally toss-ing a wool sweater into the washing machine, only to see it shrink to Barbie-size proportions. Essentially, when wool becomes warm and wet, the fibers shrink and lock together. If the garment has been ac-cidentally machine-dried as well, you probably have a lost cause on your hands. But if it's still wet and you want to try reshaping the sweater, vinegar may help: Combine the vinegar and water in a large pot and bring to a gentle boil. Add the sweater and soak in the sim-mering mixture for 10 to 20 minutes. Remove from the heat and let cool in the pot. Lay the sweater on towels to absorb the excess water, but do not rinse. Lay flat on fresh towels and gently stretch it to its original dimensions. Allow to air dry (any lingering vinegar scent will evaporate as the sweater dries).

Do You Know...

WHEN MOST VINEGAR IS PURCHASED?

Vinegar sales peak during the summer months, with the month of April following closely behind.

Personal Uses

ATHLETE'S FOOT TREATMENT

1 cup (250ml) apple cider VINEGAR
1 cup (250ml) warm water

Athlete's foot is caused by a fungus that can lead to itchy, red patches on the skin, particularly between the toes; it grows easily in warm, moist environments (like shoes and shower stalls), and the worst part is that it's easily spread, so early treatment is key. The good news is that apple cider vinegar is an effective remedy. To use, soak affected areas in a mix of equal parts vinegar and water (it's not necessary to cover the entire foot, just the inflamed parts) for 20 minutes, twice a day. Make sure to wash your feet before and after the vinegar soak, and dry thoroughly afterwards so that the fungus doesn't spread. Most cases should clear up in about 2 weeks.

> "Tart words make no friends: a spoonful of honey will catch more flies than a gallon of vinegar."
>
> –BENJAMIN FRANKLIN, AMERICAN FOUNDING FATHER (1706-1790)

WART REMOVER

1 spoonful apple cider VINEGAR

Warts can be unsightly, and sometimes painful, so many people want to get rid of them as soon as they crop up. Fortunately, apple cider vinegar has demonstrated its worth to many people seeking relief from warts. To use, soak a cotton ball in a spoonful of apple cider vinegar, place it over the wart, and use a bandage to keep it in place. Leave the cotton ball on the wart overnight. Wash the area thoroughly in the morning with soap and water, and repeat every night for up to 2 weeks.

SEE ALSO Corn Remover (lemons) on page 94.

FOOT ODOR REMEDY

2 cups (500ml) VINEGAR

If foot odor has you down, try spritzing your feet with a dose of full-strength vinegar every night for a few weeks. The vinegar will help kill odor-causing bacteria. As an added bonus, the vinegar may also help alter the pH levels on your skin, making it a less inviting place for fungus to thrive.

SEE ALSO Automatic Shoe Freshener (baking soda) on page 172 and Foot Odor Eliminator (baking soda) on page 179.

SWIMMER'S EAR PREVENTION

2 tablespoons (30ml) VINEGAR

2 tablespoons (30ml) isopropyl alcohol

Swimmer's ear is the name for a bacterial infection that occurs in the outer ear canal (versus deep inside like a normal ear infection). It commonly occurs after a day of swimming when the protective layer of wax that covers this sensitive skin can be worn away. To help get rid of bacteria-rich water in the ears after swimming, and, thus, lower the risk of an infection developing later, make a solution of equal parts vinegar and isopropyl alcohol and pack it in your beach bag. Soak a cotton ball in the mixture and squeeze a few drops into each ear after swimming.

SEE ALSO Swimmer's Hair Prevention (lemons and baking soda) on page 90.

MUSCLE ACHE RELIEVER

3 cups (750ml) apple cider VINEGAR

We often don't pay attention to our muscles until we've overworked them and start feeling some cramping and soreness set in, but you can try curbing that discomfort with a little apple cider vinegar. Add to your warm bathwater and enjoy a nice, relaxing soak.

SEE ALSO Soothing Footbath (salt and baking soda) on page 44.

HAIRBRUSH DEEP CLEANER

3 tablespoons (42g) BAKING SODA

3 tablespoons (45ml) VINEGAR

Warm water

Hairbrushes and combs need a deep clean every now and then to remove the buildup of styling products and oil that can accumulate; otherwise, a dirty brush can make for dirtier hair as you inadvertently put small amounts of those materials back into your hair every time you use it! To get your beauty tools spic-and-span the all-natural way, start by placing them in the sink and sprinkling them with the baking soda. Pour the vinegar over them and let them bubble and fizz for a few minutes. Fill the sink with warm water and let the brushes soak for a few hours. Rinse thoroughly, taking care to remove any trapped hair, and allow to air-dry.

Do You Know...

WHY VINEGAR IS PASTEURIZED?

Vinegar is filtered and pasteurized to make it crystal clear. Better-quality aged wine, cider, or malt vinegars are often left unfiltered and non-pasteurized, in which case the bacteria will form at the top and sometimes sink to the bottom, rendering them cloudy. NOTE: The bacteria in non-pasteurized vinegar is not harmful, and some natural food enthusiasts believe it even possesses health benefits.

HERBAL HAIR RINSE

1 cup (250ml) VINEGAR
½ cup (20g) fresh lavender
2 cups (500ml) water

Do you want soft, lustrous hair but cringe at the thought of dousing your head with vinegar? Then make your own herbal version instead. Combine the vinegar with the lavender flowers (or rose petals, or whatever fragrant flower is growing in your garden) in a glass jar. Let the mixture sit, covered, for about 2 weeks, and then strain through a fine-mesh sieve. Add the water to dilute the vinegar and pour into a plastic bottle. Pour a handful over your hair after shampooing and rinse. Your locks will feel amazing.

SEE ALSO Hair Revitalizer (salt) on page 45; Hair-Clarifying Rinse (lemons and salt) on page 92; and What Else Can It Do? (baking soda) on page 176.

> "To make a good salad is to be a brilliant diplomatist—the problem is entirely the same in both cases. To know exactly how much oil one must put with one's vinegar."
>
> —OSCAR WILDE, ANGLO-IRISH WRITER (1854-1900)

Household Uses

PLASTIC SHOWER CURTAIN CLEANER

Laundry detergent

1 cup (250ml) VINEGAR

Soap scum and soap film on a plastic shower curtain can make it look dingy and old. Plus, there's the problem of mold and mildew that can accumulate. To remove soap scum and keep mold problems at bay, place the shower curtain in the washing machine with a few large towels. Add detergent and the vinegar and wash on the gentle setting. Remove the curtain right after the spin cycle, and hang it back in the shower. TIP: To prevent future buildup after laundering, dip a clean washcloth in a cup (250ml) of vinegar, coat the curtain with a final dose of vinegar, and allow to air-dry.

SEE ALSO Mildew Stain Remover (lemons and salt) on page 84 and What Else Can It Do? (baking soda) on page 181.

SHOWERHEAD LIME-DEPOSIT REMOVER

1 cup (250ml) VINEGAR

Over time, mineral deposits can block the small nozzles of your showerhead and affect water pressure, but a 12-hour soak in vinegar can do wonders to get rid of them. Soak a clean washcloth in 1 cup (250ml) vinegar and wrap it around the showerhead. Alternately, you can give the fixture a direct soaking by filling a 1-gallon plastic bag with vinegar and affixing it to the fixture with some duct tape (to hold it in place). Let sit overnight and then remove. Your morning shower should be noticeably improved.

SINK AND COUNTERTOP DISINFECTANT

1/4 cup (60ml) VINEGAR

Bacteria are everywhere, but you can minimize the presence of the harmful varieties on sinks and countertops with a good dose of vinegar—its acidity not only kills most bacteria but can also inhibit future growth. Use a clean cloth to apply 1/4 cup (60ml) vinegar to countertops and sinks. Do not rinse with water, but, instead, allow to air-dry (the sour scent will fade as it dries).

SEE ALSO What Else Can They Do? (lemons) on page 101 and Super Simple Sink Cleaner (baking soda, lemons, and vinegar) on page 167.

WINDOW AND MIRROR CLEANER

¹/₂ cup (125ml) VINEGAR
Newspaper

Want a streak-free way to clean mirrors and windows? Pour about ¹/₂ cup (125ml) vinegar into a shallow bowl or pan. Crumple a sheet of newspaper and dip into the vinegar. Wipe the glass several times with the newspaper until the surface is almost dry. Then shine with a clean, soft cloth or dry newspaper. Wear gloves if you don't want to get newsprint ink on your hands. TIP: If there are streaks left on your window after using this cleaning method, it's most likely due to a waxy residue left from prior use of a commercial cleanser. Add a dab of dish soap to the vinegar to help eliminate this buildup.

SEE ALSO Window and Mirror Cleaner (lemons) on page 100.

BATHROOM FIXTURE CLEANER

³/₄ cup (175ml) VINEGAR
³/₄ cup (175ml) water

Owing to its grease-fighting ability, vinegar is an excellent resource for polishing chrome and stainless steel surfaces like bathroom fixtures. Keep a spray bottle filled with equal parts vinegar and water with your bathroom cleaning supplies (change the amount if necessary, but keep the ratio 1:1). Just a quick squirt on the fixtures is all you need to keep them spotless. Wipe dry and buff to a shine with a dry paper towel.

ALL-PURPOSE FLOOR CLEANER

1 cup (250ml) VINEGAR

1 cup (250ml) isopropyl alcohol

1 cup (250ml) water

6 drops dishwashing liquid

For fast floor cleanups, keep a bottle of this all-purpose cleaner handy: Fill a medium spray bottle with the vinegar, alcohol, water, and dishwashing liquid. Shake gently to combine. To use, spray sparingly and mop until dry.

WOOD AND LAMINATE FLOOR CLEANER

1/2 to 1 cup (125 to 250ml) VINEGAR

1 gallon (4 liters) warm water

Keeping wood floors clean is the key to keeping them lustrous and beautiful. After a thorough sweeping, pick up any residual dirt with a damp mop and a combination of 1/2 cup (125ml) vinegar and the water (if you have been using oil soap on your floor, increase the vinegar to 1 cup or 250ml). The vinegar will prevent hard water spots from forming on the floor as it dries.

WOOD PANELING POLISH

2 tablespoons (30ml) olive oil

1/4 cup (60ml) VINEGAR

2 cups (500ml) warm water

You can restore the warm, rich glow of most wood paneling with this simple formula: Combine the olive oil, vinegar, and warm water in a wide-mouth jar with a lid (an empty mayonnaise container works well). Shake until well combined. Wipe evenly on paneling with a clean cloth and let dry.

EASY WOODWORK, WALL, AND WINDOW BLIND CLEANER

1 cup (250ml) ammonia

1/2 cup (125ml) VINEGAR

1/4 cup (55g) BAKING SODA

1 gallon (4 liters) warm water

Make spring cleaning a little easier by washing painted walls, woodwork, and Venetian blinds with this simple formula: Combine the ammonia, vinegar, and baking soda in a large bucket with the warm water. Use a clean sponge or cloth to wipe this solution over walls or blinds, and then rinse thoroughly with clear water. Dirt and grime will come off easily without dulling the painted finish or leaving streak marks.

SEE ALSO Window Blind Cleaner (vinegar) on page 141.

WINDOW BLIND CLEANER

1 cup (250ml) VINEGAR
1 cup (250ml) warm water

Before tossing out your old cotton socks that have lost their mates, consider saving a few for spring cleaning—you can put them to good use by getting your window blinds nice and clean. Just put an old sock on your hand as if it were a mitten and dip into a bowl containing equal parts vinegar and warm water. Squeeze dry and run the dampened sock along the slats to remove dust. Repeat as necessary until clean.

SEE ALSO Easy Woodwork, Wall, and Window Blind Cleaner (vinegar and baking soda) on page 140.

CRYSTAL CHANDELIER CLEANER

½ cup (125ml) VINEGAR
½ cup (125ml) water

Cleaning a chandelier can seem daunting, but it's a task that should be undertaken every few years. It's a chore that's well worth the time and effort once you see the brilliant shine restored. Before you begin, turn the lights off and give the bulbs time to cool (do not turn it back on during cleaning). Meanwhile, combine equal parts vinegar and water in a small spray bottle. Use a cloth to remove most of the surface dirt, then spray a clean, lint-free cloth with the vinegar mixture and polish each piece of crystal (spraying the chandelier directly can damage the metal fixtures).

LEATHER CONDITIONER

1 tablespoon (15ml) VINEGAR

2 tablespoons (30ml) linseed oil

To clean and condition most leather items (except shoes, which are often made of a harder leather), use 1 part vinegar with 2 parts linseed oil. Mix the two together in a small glass container. Apply to leather furniture with a soft cloth and buff lightly. NOTE: Rags soaked in linseed oil are considered a fire hazard, so immediately soak them in water after use and never put them in the dryer.

SEE ALSO Leather Spot Cleaner (lemons) on page 84.

PAINTBRUSH SOFTENER

2 to 3 cups (500 to 750ml) VINEGAR

No matter how well you wash out a paintbrush, odds are it will have stiffened by the next time you want to use it. Of course, vinegar offers an excellent means of softening and restoring paintbrushes. Place the brushes in a small, heavy-bottomed pot (one that you can dedicate to household situations and not use for cooking) and pour enough vinegar (2 to 3 cups, or 500 to 750ml) over the brushes to soak the bristles. Let sit for 1 to 2 hours. Then place the pot over medium heat and simmer for 10 to 20 minutes. Rinse with cool water, and you're ready to paint!

WOOD GLUE SOFTENER

1 tablespoon (15ml) VINEGAR

When repairing antique furniture or other wooden items, it's important to get rid of old glue beforehand to ensure the best bond possible. To soften the glue and make scraping it away much easier, use
about 1 tablespoon (15ml) vinegar. For better access to wooden joints,
fill a small medicine dropper with vinegar (the plunge-style dropper
made for infants works well) and use it to squirt a small amount right
where you need it. Naturally, vinegar also works well on cleaning up
glue spills. NOTE: Vinegar may not be effective on more modern furniture that has been assembled with epoxy glues.

BABY TOY DISINFECTANT

1/4 cup (60ml) VINEGAR

Since babies are especially prone to putting things in their mouths,
you naturally want to keep your little one's toys as clean as possible.
There's really no need to rely on harsh disinfectants when you have
a bottle of vinegar in the house. Just soak a soft clean cloth in 1/4 cup
(60ml) vinegar and use it to wipe down all the non-plush toys. Allow
items to air dry (the vinegar scent will fade as it dries).

SEE ALSO Stuffed Animal Cleaner (baking soda) on page 184.

URINE SMELL ELIMINATOR

2 cups (500ml) VINEGAR

1 cup (250ml) water

1 cup (220g) BAKING SODA

Repeated bedwetting not only makes for unpleasant sleep, but also a ruined mattress. Fortunately, you can rely on vinegar to get rid of the smell of urine. Mix the vinegar and water and apply with a sponge to affected areas. Blot dry with a towel and cover the spot with the baking soda. Remove the baking soda with a handheld vacuum when it is dry to the touch.

What Else Can It Do?

CLEAN BALLPOINT PEN MARKS

Although inexpensive, reliable, and virtually maintenance-free—compared to fountain pens—a trusty ballpoint pen is still capable of making quite a mess when the ink lands where you'd rather it not. To get rid of errant ink marks on walls and clothing, dip a clean cloth or sponge in a small amount of vinegar and apply sparingly to the stain until it is lifted.

OLD WALLPAPER PEELER

1 cup (250ml) VINEGAR
1 quart (1 liter) warm water
2 to 3 drops dishwashing liquid

Need to take down some wallpaper? Before you rush out to rent a cumbersome steam machine or buy an expensive paper-removing product, whip up a batch of this vinegar formula for a mere fraction of the cost: In a large bucket or bowl, combine the vinegar and warm water and add a few drops of dishwashing detergent to the mix (feel free to make a larger batch if necessary). Use a sponge to apply the solution liberally to the old paper. Let stand for 5 to 10 minutes and then scrape off the paper with a putty knife. To remove all traces of glue, use a nylon sponge to give the walls a final wash with the same solution.

LONG-LIFE FLOWER WATER

1 quart (1 liter) warm water
3 tablespoons (38g) sugar
2 tablespoons (30ml) VINEGAR

Make fresh-cut flowers last as long as you can by placing them in the best possible water. For every quart (liter) of warm water, add 3 tablespoons (38g) sugar and 2 tablespoons (30ml) vinegar. Mix well to ensure the sugar is dissolved and use to fill your flower vase. The arrangement will stay fresher longer because the sugar helps nourish the flowers and the vinegar inhibits bacteria growth in the water.

Outdoor Uses

DEER, RACCOON, AND OCCASIONAL PICNIC PEST DETERRENT

1 quart (1 liter) VINEGAR

For all of vinegar's many uses, the animal kingdom appears not to have caught on to its alluring power. In fact, most creatures seem to despise it. So turn that natural aversion to your advantage, and let a bit of vinegar keep unwanted critters at bay. Decorate your garden with rags that have been soaked in vinegar to keep deer from munching your vegetables; likewise, if you have more full trash bags than you have containers, discourage animals from getting too curious by dousing the outside of each bag with a liberal splash of vinegar (don't count on this to work, however, if you live in bear country). Even flying pests will be perturbed by the smell of vinegar: Place a few small bowls of vinegar in strategic spots at your next picnic (e.g., near the dessert table), and flies will choose a different dining spot.

JELLYFISH STING SOOTHER

1 cup (250ml) VINEGAR

For a day at the beach, make sure to pack a bottle of vinegar in your beach bag—because nothing can ruin a perfect day like a jellyfish sting. The long tentacles of the jellyfish release a painful toxin, so you are sure to feel the pain as soon as you've been stung. Whatever you do, don't pour fresh water on the sting, as the change in pH levels can actually cause the jellyfish to pump out even more venom. Instead, douse the affected area with about 1 cup (250ml) vinegar to neutralize the poison, and then carefully remove the tentacles. Seek immediate medical treatment if you develop any signs of muscle pain, shortness of breath, backache, or hives, as these can be a sign of a sting from a more dangerous sea creature—the Portuguese man-of-war.

SEED GERMINATOR

1 teaspoon (5ml) VINEGAR
3 tablespoons (45ml) water

Most seeds require damp conditions for optimal germination; unfortunately, the same conditions are also ideal for a variety of molds that can threaten a young seedling. If you start to see mold spores collecting, take matters into your own hands and fill a small spray bottle with the vinegar and water. Spray seedlings lightly to clean them off, and then transfer to a new container. Spritz seeds regularly with this diluted mixture while you wait for them to sprout.

What Else Can It Do?

ELIMINATE WEEDS

If that dandelion patch has grown wildly out of control, grab a bottle of vinegar and pour 3 tablespoons (45ml) directly on the base of each flower. The acidity will make the soil far less inviting. But apply carefully because vinegar can do a good job of killing the grass you'd like to keep, too. You can also apply undiluted vinegar to plants that grow between the cracks in a driveway or sidewalk.

SEE ALSO Poison Ivy Killer (salt) on page 58 and What Else Can It Do? (baking soda) on page 189.

BUMPER STICKER REMOVER

1/2 cup (125ml) VINEGAR

While bumper stickers are an incredibly popular means of self-expression, they can become a hindrance if you want to sell your car, and their adhesives make them notoriously difficult to remove—unless you know the vinegar secret. Soak a sponge in 1/2 cup (125ml) vinegar and use it to soak the bumper sticker thoroughly. Wait 30 minutes and the edges of the sticker should start to come undone. Work slowly and carefully to pull the sticker away, applying more vinegar as necessary to dissolve the glue. Rinse thoroughly with fresh water when you're done.

WINDSHIELD WIPER-BLADE CLEANER

¼ cup (60ml) VINEGAR

If your windshield wiper blades seem to skip or squeak, it could be that dirt is to blame. One of the best ways to clean them is with a soft cloth soaked in ¼ cup (60ml) vinegar. Wipe down each blade from top to bottom and then run the wipers with a few squirts of windshield cleaning fluid to clear the vinegar away.

What Else Can It Do?

KEEP BUGS AT BAY

Keeping the surface of your skin slightly acidic will make you less delectable to flying insects and their painful, irritating bites. Vinegar can do the job and is much less harsh than many commercial insect repellents. To use, pour a spoonful of white or cider vinegar on a sheet of paper towel and rub lightly over exposed skin. Vinegar is also used as a skin tonic, so it should feel very refreshing. (The sour smell should fade as it dries.)

SEE ALSO Mosquito Repellent (lemons) on page 106 and Bug Bite Remedy (baking soda) on page 175.

EGGED CAR CLEANER

¼ cup (60ml) VINEGAR
1 gallon (4 liters) warm water

Most likely a result of a practical joke, an egg-splattered car is not a laughing matter. Eggs can leave a protein-based residue that can badly damage a car's paint surface. To remedy the situation, act as quickly as possible: Mix the vinegar into the warm water and transfer to a spray bottle. Apply the vinegar solution to the affected areas liberally, picking off as much of the eggshell as possible (the shell can scratch the paint, so remove all that you can beforehand). Let sit for 5 to 10 minutes and then use a clean, damp rag to wipe the car clean. Repeat, as necessary, until all egg is removed.

What Else Can It Do?

SHINE YOUR CAR

Give the shiny metal trim on your car the royal treatment with a clean, soft cloth dipped in vinegar (never use abrasive cleansers as they will scratch and dull the finish).

Pet Uses

CAT BEHAVIOR MODERATOR

½ cup (125ml) VINEGAR
½ cup (125ml) water

If your cats have a bad habit of fighting with one another or scratching upholstery, a bit of vinegar might be all you need to correct unwanted behavior. Fill a spray bottle with equal parts vinegar and water. If you come upon your cats fighting one another, a squirt in their general direction should be enough to get them to go their separate ways. If it's furniture that you need to protect, test the solution first in an inconspicuous place. If the vinegar doesn't affect the color, go ahead and lightly spray the spots you'd rather your cats not claw—the smell will keep them away (the scent will be largely undetectable to humans once it dries).

SEE ALSO Cat Deterrent (lemons) on page 108.

"The star of oil and vinegar and the oil and vinegar of the stars."

–PAUL NEWMAN, AMERICAN ACTOR AND PHILANTHROPIST (1925-2008)

CANINE EAR CLEANER

1 tablespoon (15ml) VINEGAR

1 tablespoon (15ml) water

1 tablespoon (15ml) hydrogen peroxide

Long-eared dogs tend to be more vulnerable to ear problems because the floppy nature of their ears makes a ripe breeding ground for moisture-loving yeast and bacteria. If your pet is prone to such problems, help keep Fido's ears clean with a gentle vinegar formula. Combine equal amounts of vinegar, water, and hydrogen peroxide in a small plastic bowl. Dip a cotton ball or a finger wrapped in gauze into the vinegar solution and wipe carefully along the visible parts of your dog's inner ears (never use cotton swabs, which could puncture an eardrum).

What Else Can It Do?

PREVENT FUTURE PET ACCIDENTS

As any dog or cat owner will tell you, pets have a frustrating tendency to return to the scene of the crime and soil the same section of carpet over and over again. To stop the cycle, it's important to neutralize the smell, and vinegar can do the trick. Simply blot up as much of the mess as possible and clean with soap and water as you normally would. Then spray the area with vinegar and let air dry. The offensive odors will be gone and the vinegar scent will fade considerably.

Kids' Activities

TURN A CHICKEN BONE INTO RUBBER

1 cup (250ml) VINEGAR

The next time you roast a chicken, set aside a thigh bone. Let the kids soak it in 1 cup (250ml) vinegar for 3 or 4 days. When they take the bone out, they'll be amazed to see how easily it can be bent—the vinegar leaches some of the calcium, leaving the bone noticeably softer and more pliable.

MAKE A SEE-THROUGH EGG

1 egg
1 small plastic bowl
VINEGAR

Place an egg in a small plastic bowl and cover completely with vinegar. Refrigerate for 24 hours and then use a slotted spoon to lift the egg gently from its vinegar bath. Carefully rinse with water and repeat the process if the shell still remains. Cover with a fresh batch of vinegar and refrigerate for another 24 hours. Soon enough, you should be able to see the inside of the egg through its translucent and flexible membrane.

CHAPTER 4

Baking Soda

♦ ♦ ♦

Known by the technical name sodium bicarbonate, or sometimes bicarbonate of soda, baking soda, is a white crystalline substance known chemically as $NaHCO_3$. Although it is found in nature as a component of the minerals natron and trona, it is usually manufactured using the Solvay process, which transforms the readily available substances, salt brine and limestone, into baking soda. Three quarters of the annual production of baking soda uses this process. The human body is also one of the best manufacturers of sodium bicarbonate, as it plays an important role in protecting your teeth by neutralizing acids in saliva, as well as preventing ulcers (it neutralizes stomach acids, too).

When baking soda comes into contact with acids, it effervesces, producing carbon dioxide, and it is alkaline (it has a pH of 8.1, while 7 is neutral).

A Brief History of Baking Soda

Baking soda is actually the one kitchen staple in this book with a relatively modern history. The story begins just a few hundred years ago and starts, of course, with baking.

Until the 1760s yeast was the primary ingredient used to help baked goods rise. Another method involved mixing or kneading extensively to trap small air bubbles in the batter or dough before baking. Both methods, however, were slow and time-consuming, so when American colonialists discovered the leavening powers of potash (a potassium-carbonate made from boiling hardwood ashes until they're dry, originally used by soap and glass manufacturers), the race was on to master a way to use it. One of the challenges that potash presented was that it didn't always produce consistent results. Sometimes the fizzy chemical reaction responsible for leavening would end before the baker could get the batter in the oven. Other times, it would lend an unpleasant taste to the finished product.

In 1791 Nicolas Leblanc successfully developed a process that converted salt into soda ash (a sodium carbonate). Soda ash then replaced potash for baking purposes. Finally, around 1839, Dr. Austin Church developed a process that could effectively transform soda ash into baking soda; better yet, Church's baking soda could be produced more cheaply than potash or soda ash and proved to be much more reliable. The company he started eventually became Church and Dwight, and while the soda ash process was later replaced by mining a mineral called trona to produce sodium bicarbonate, the same business has been making baking soda under the Arm & Hammer brand for over 160 years.

Do You Know...

HOW TO TEST BAKING SODA'S EFFECTIVENESS?

Mix a $1/4$ teaspoon (1g) baking soda with 2 teaspoons (10ml) vinegar; if you don't immediately see bubbles, the baking soda is no longer fresh and should be replaced with a new box.

How Baking Soda Works

To understand the mystery of baking soda, let's start with a brief review of the basic chemistry behind its magic. Many stains are acidic in nature, and the opposite of acidic is alkaline, and that's where baking soda comes in. When baking soda is added to a mildly acidic substance, such as lemon or vinegar, it reacts by neutralizing the acid. In doing so, it produces lots of bubbles, which are mostly small amounts of carbon dioxide gas.

In a recipe, baking soda has to be paired with other acidic ingredients (e.g., lemon juice, vinegar, buttermilk, yogurt) to stimulate the leavening process. Essentially, the small bubbles lift the starches present in the batter or dough, and then the heat of baking or

"It's filled with . . . baking soda. Because it really smells."

–KATE O'BRIEN, IRISH NOVELIST AND PLAYWRIGHT, (1897-1974)

frying sets them in place. Just a pinch of baking soda also helps vegetables maintain their color while cooking, but this use fell out of favor when it was discovered that vitamins were destroyed in the process.

In terms of household use, baking soda is quite versatile for its abrasive abilities and stain-lifting power. Its fine grain makes for an exceptional abrasive that can scour without scratching, and its ability to neutralize acids translates to a variety of household uses, from maintaining a proper pH balance in a swimming pool to eliminating odors in the refrigerator. When dissolved in water, baking soda helps detergents work better because it maintains a neutral pH in the water, which is optimal for cleaning. Plus, since it's able to neutralize both acids and bases, baking soda actually eliminates odors rather than just covering them up.

Do You Know...

WHY HOUSEHOLD BAKING SODA SALES ARE NOT AS SIGNIFICANT AS THEY WERE IN THE EARLY 1900S?

In the last century, the introduction of self-rising flour and baking mixes has decreased its demand as a necessity for baking.

Kitchen Uses

REFRIGERATOR ODOR EATER

1 16-ounce box (454g) BAKING SODA

Baking soda's use as a refrigerator deodorizer is probably most familiar. Its deodorizing properties can do wonders to control food odors. Open the top of a box and place it in the back of the refrigerator or freezer. Replace every 1 to 2 months for best results, but don't throw the old box away—reuse the baking soda to clear the pipes for the Drain Clearer on page 182.

SEE ALSO Refrigerator Cleaner (salt and baking soda) on page 31.

What Else Can It Do?

AVOID DISHWASHER HANDS

When wearing rubber gloves to wash dishes or undertake any other cleaning chore that may get your hands wet, sprinkle some baking soda into the gloves to make it easier for you to slide the gloves on and off. The baking soda will also help your hands stay drier, and the gloves will also smell fresher longer.

GARBAGE CAN DEODORIZER

¼ cup (55g) BAKING SODA

Looking for a way to control garbage odors but dislike the overpowering smell of scented trash bags? Let a dash of baking soda neutralize the bad odors for you. Just get in the habit of pouring ¼ cup (55g) baking soda into your kitchen garbage bags before you put them in the pail, and the baking soda will help absorb unpleasant trash odors.

FRUIT AND VEGETABLE BATH

1 gallon (4 liters) water
¼ cup (55g) BAKING SODA

When you come home from the farmers' market and have a basket full of produce to wash before storing, consider a baking soda bath to make your task easier. Combine the water and baking soda in the sink. Add your vegetables, swish them around, and let them have a brief soak. To remove the bitter taste of baking soda, make sure to follow up with a thorough rinse.

SEE ALSO Fruit and Vegetable Wash (lemons and baking soda) on page 74.

FRUIT AND VEGETABLE SHELF-LIFE EXTENDER

¼ cup (55g) BAKING SODA

2 paper towels

Humidity is the bane of your refrigerator's crisper drawer. Too much moisture provides an opportunity for microscopic spores on fruits and vegetables to grow more quickly, a problem that can ultimately result in spoiled produce. To help keep conditions in your crisper on the dry side, stack 2 paper towels in the bottom of the drawer and sprinkle ¼ cup (55g) baking soda in between them. Replace every week or so when you replenish your produce.

Do You Know...

THE DIFFERENCE BETWEEN BAKING SODA AND BAKING POWDER?

It's a question that confuses many cooks, but it's important to remember that they work very differently in a recipe, so don't substitute one for the other. They're both chemical leavening agents that cause batter and dough to rise and then set when heated in the oven. So what's the difference? Baking soda is four times as strong as baking powder and requires the presence of an acid in the recipe to initiate the leavening process. Baking powder contains an acid, so other acidic ingredients are not needed in the recipe in order for it to do its job.

STICKY MESS CLEANER

3 spoonfuls BAKING SODA

Reorganizing the kitchen cupboards can be a gratifying task until you come across a sticky mess stuck to a shelf. Before you haul out the scrub brush and a slew of cleansers, consider sprinkling a few spoonfuls of baking soda on the mess. Owing to the delicate scrubbing power of baking soda, you should be able to get away with just a damp sponge treatment. Make sure to rinse thoroughly with fresh water.

CLEANER FOR MOST POTS AND PANS

¼ cup (55g) BAKING SODA
2 to 3 cups (500 to 750ml) water

If you have a dirty pot to contend with, reach for the baking soda. Sprinkle the baking soda on the bottom and add the water. Bring to a boil and simmer for a few minutes; the baking soda will significantly loosen the mess. Let cool, and your scrubbing job will be much easier. Don't, however, try this formula on aluminum pots (including the popular anodized aluminum types), as baking soda will react with the metal and could damage the finish.

SEE ALSO All-Purpose Metal Cookware Scrub (vinegar and salt) on page 121.

What Else Can It Do?

MAKE BAKING POWDER

Because baking soda and baking powder can react very differently in a recipe, you shouldn't consider them interchangeable. But if you find yourself short a little baking powder, you can make due with baking soda and the following formula: For every teaspoon (5g) of baking powder, use a combination of 1/4 teaspoon (1g) baking soda, 1/2 teaspoon (1.5g) cream of tartar, and 1/4 teaspoon (1g) cornstarch. Make each batch as needed.

EASY SILVER CLEANER

2 quarts (2 liters) hot water

1 cup (220g) BAKING SODA

Aluminum foil

If you've ever spent hours polishing a set of silverware, you'll be glad to know there's an easier, greener way to get the job done. Pour the hot water into a gallon-size plastic bucket or large plastic dishpan, add the baking soda, and stir to dissolve. Place a strip of aluminum foil about 2 inches wide in the pan along with your silver items and let everything soak for 30 minutes. The tarnish will lift off your silver and attach itself to the aluminum. Remove all the now-shiny silver items, rinse thoroughly with fresh water, and dry with a soft cloth.

KITCHEN SPONGE REVIVER

2 tablespoons (28g) BAKING SODA
2 tablespoons (30ml) VINEGAR
Water

Few things can smell as foul as kitchen sponges past their prime, owing to the bacteria they can collect. It's best to replace them frequently to maintain the cleanest possible kitchen; however, that is not to say the sponges need to be disposed of completely. They can still come in handy for a variety of projects around the house, if you revive them with a healthy dose of baking soda and vinegar. Mix equal amounts on a sponge and add enough water to saturate. Squeeze gently to ensure the mixture has worked all the way through the sponge. Let sit for about an hour, and then wring dry. Repeat if necessary.

STAINLESS STEEL APPLIANCE CLEANER

1 small spoonful BAKING SODA

Stainless steel appliances are popular in part because they tolerate heat well, resist germs, and look great, but they do require careful cleaning. For most messes, baking soda is the perfect cleaner for the job because it won't scratch the metal. To use, put a small spoonful on a damp sponge and squeeze the sponge a few times to form a thick paste. Rub lightly on stainless steel in the direction of the grain. Make sure to rinse thoroughly and buff with a dry paper towel to remove water spots.

SUPER SIMPLE SINK CLEANER

1 tablespoon (14g) BAKING SODA

1 teaspoon (5ml) LEMON juice

2 tablespoons (30ml) VINEGAR (optional)

It's easy to keep porcelain sinks sparkling clean with this all-natural, germ-fighting combination: To begin, place the baking soda in a small pile in the sink. Pour the lemon juice on top and watch it start to bubble. Stir with your finger to form a paste, then lightly rub the paste over the entire sink with a damp sponge. Rinse thoroughly with fresh water. For an extra sanitizing measure, pour the vinegar on the sponge and use it to give everything one last wipe down. (The vinegar scent will fade as it dries).

SEE ALSO Sink and Countertop Disinfectant (vinegar) on page 137.

Do You Know...

WHICH CAKE RECIPE IS KNOWN FOR ITS USE OF BAKING SODA?
Baking soda causes reddening of cocoa powder when baked, hence the name, Devil's Food Cake.

STOVETOP FIRE FIGHTER

1 cup (220g) baking soda

Keep a box of baking soda in the cabinet nearest the stove in case your pan has a flare-up during cooking. To use, quickly cover the bottom of the pan with a layer of baking soda (about 1 cup or 220g) and place a lid on top. Do not try to wash out the pan until it is completely cool, and never pour water on a grease fire.

SEE ALSO Fire Fighter (salt) on page 29.

What Else Can It Do?

KEEP CANDY FROM CRYSTALLIZING

When making caramel corn and other types of homemade candy that require a sugary syrup base, a pinch of baking soda added at the end of the boiling step will keep any undissolved sugar crystals from ruining the batch.

Laundry Uses

LAUNDRY DETERGENT BOOSTER

½ cup (110g) baking soda

Do you want whiter whites, brighter brights, and all of the wonderful benefits that laundry detergents usually promise? Well, if you've purchased a brand that isn't living up to your expectations, baking soda can save the day. Add ½ cup (110g) to the wash cycle and your clothes really will come out a little perkier. If your clothes aren't particularly soiled, you may even be able to use less detergent with the baking soda boost because baking soda maintains a neutral pH level in the water, enabling the detergent to perform at its absolute best. TIP: Baking soda works best with liquid detergents as opposed to powdered.

SEE ALSO All-Natural Fabric Softener (lemons and baking soda) on page 82 and What Else Can It Do? (vinegar) on page 126.

PRE-TREATMENT FOR KID STAINS

1 tablespoon (14g) BAKING SODA

1 teaspoon (5ml) VINEGAR

Ridding children's clothes of grass stains and crayon marks can feel like an uphill battle, unless you decide to rely on a trusty baking soda and vinegar combo to help lift the stains. To use, mix a paste of the baking soda and vinegar. When the fizzing subsides, dip an old toothbrush in the mixture and use to gently work the paste into the stain. Let the mixture sit for a few minutes and then launder as usual.

SEE ALSO Pre-Treatment for Grass Stains (lemons and baking soda) on page 86 and What Else Can It Do? (baking soda) on page 184.

What Else Can It Do?

REFRESH CLOSETS AND GARMENT BAGS

Closets and garment bags aren't afforded a lot of air circulation, so it's common for them to develop stale odors over time. To help keep closets smelling fresh, place a box of baking soda on the floor and replace every couple of months. When storing clothes in garment bags, put 1/4 cup (55g) baking soda in the bottom of a clean cotton sock and tie the end. Place the sock in the bottom of the bag to absorb any odors that might develop.

SEE ALSO Lemon Sachet (lemons and baking soda) on page 98.

BABY CLOTHES AND CLOTH DIAPER CLEANER

$^{1}/_{2}$ cup (110g) BAKING SODA

Fragrance-free liquid laundry detergent

$^{1}/_{4}$ cup (60ml) VINEGAR

Baking soda is a great laundry aid for baby clothes because it helps neutralize the strong acids present in the bodily fluids of babies. Plus, being fragrance-free and water-soluble, it dissolves completely in the laundry water and doesn't leave behind harsh chemicals that can irritate a baby's delicate skin. To use, add the baking soda to the wash cycle along with the detergent. Add the vinegar to the rinse cycle to leave clothes naturally softened and ensure that all the detergent is properly removed.

SEE ALSO Baby Formula Stain Remover (lemons) on page 86 and Diaper Rash Soother (baking soda) on page 178.

LAUNDRY ODOR TAMER

1 16-ounce box (454g) BAKING SODA

If your busy schedule doesn't afford enough room for tackling laundry during the week, you might start noticing the scent of exercise clothes and other well-worn garments long before you're able to wash them. To minimize sour laundry smells, keep a box of baking soda near the hamper. Add a small handful whenever a potentially pungent item is added to the mix. The baking soda will also boost the effectiveness of your detergent in the wash after neutralizing the smells.

AUTOMATIC SHOE FRESHENER

1 cup (220g) BAKING SODA, divided

If you or your loved ones are bothered by the smell of shoes after you've worn them, here's a great method for getting rid of the offensive odors. Take an old pair of clean cotton socks (no holes in the toes, please) and fill each one with $1/2$ cup (110g) baking soda. Shake gently to make sure most of the powder is near the toe of the sock, and tie a knot to secure the powder inside. Place one sock in each shoe at the end of the day, and the baking soda will help absorb moisture and neutralize the odors. Replace every month.

SEE ALSO Foot Odor Remedy (vinegar) on page 132 and Foot Odor Eliminator (baking soda) on page 179.

Do You Know...

WHERE BAKING SODA COMES FROM?

The largest known deposit of trona is located in the United States in Green River, Wyoming, where the remains of an ancient lake left a rich supply.

Personal Uses

CANKER SORE RELIEF

½ teaspoon (2.5g) BAKING SODA

1 small glass warm water

Those painful white canker sores that occasionally pop up in your mouth can make it difficult to eat, so they demand swift treatment. One common formula is to make a mouth rinse out of ½ teaspoon (2.5g) baking soda and a small glass of warm water. Stir to dissolve the baking soda completely and then take a sip to swish around your mouth (do not swallow). Repeat 3 or 4 times a day until the sore is gone.

What Else Can It Do?

CLEAR A STUFFY NOSE

If cold symptoms have made it hard to breathe through your nose, try clearing the congestion by mixing ¼ teaspoon (1g) baking soda into 1 tablespoon (15ml) water. Soak up the mixture with a cotton ball and then gently squeeze a few drops into each nostril.

TOOTH-WHITENING TOOTHPASTE

$\frac{1}{2}$ teaspoon (2.5g) BAKING SODA

$\frac{1}{2}$ teaspoon (3ml) water

Baking soda is great at polishing and whitening your teeth. To use, mix $\frac{1}{2}$ teaspoon (2.5g) with a few drops of water in the palm of your hand to make a paste. Dip your toothbrush in the mixture and brush as usual; rinse thoroughly when you're done. NOTE: Baking soda does contain quite a bit of sodium, so do not use if you've been diagnosed with gum disease or if you're on a salt-restricted diet.

What Else Can It Do?

CLEAN YOUR TOOTHBRUSH

Perhaps it sounds odd to talk about taking some time to clean the tool that cleans your teeth, but stop and think about it: If you are using that little brush 2 or 3 times a day, what exactly happens to the mix of toothpaste and saliva left on it afterwards? No matter how well you rinse the brush after brushing, there's a good chance a layer of toothpaste residue will accumulate long before it is ready to be replaced. So make it a point to put 1 teaspoon (5g) baking soda in the bottom of your rinsing glass, fill it with water, and soak the bristles of your toothbrush for an hour or so every week. It will loosen all the toothpaste residue and could make your brush last just a little bit longer.

BUG BITE REMEDY

1 spoonful BAKING SODA

1 spoonful water

To relieve the itch and pain that accompanies mosquito bites, turn to baking soda for quick relief. Mix a spoonful of baking soda with enough water to make a thick paste. Dab the mixture on affected areas and let dry (rinse off only when necessary). Repeat as needed.

SEE ALSO Mosquito Repellent (lemons) on page 106 and What Else Can It Do? (vinegar) on page 149.

GENTLE FACIAL SCRUB

1 small spoonful BAKING SODA

1 quarter-size dollop facial cleanser

Sloughing away dead skin cells is an excellent way to polish the skin and reveal a brighter, smoother complexion. If you're prone to dry skin, try boosting the scrubbing capacity of your favorite face wash once a week: Mix the usual amount of cleanser with a small spoonful of baking soda in the palm of your hand. Wash as usual, applying very little pressure as you let your fingertips move in small circular patterns. The baking soda will gently scrub your skin. Make sure to rinse thoroughly and pat dry (do not use on broken or irritated skin).

SEE ALSO Moisturizing Body and Facial Scrub (salt) on page 44.

HOME MANICURE OR PEDICURE SOAK

¼ cup (55g) BAKING SODA

2 quarts (2 liters) warm water

If you have dry cuticles and rough spots, take a few minutes to treat your hands and feet to the royal treatment by first soaking them in a warm baking soda solution for 10 minutes. After you're done soaking, tackle calluses with a pumice stone and push back your cuticles. And don't forget to moisturize when you're through!

SEE ALSO Soothing Footbath (salt and baking soda) on page 44 and Nail Brightener (lemons and vinegar) on page 95.

What Else Can It Do?

GIVE YOUR HAIR A FRESH START

Using a variety of hair products day in and day out can, over time, start to weigh down your locks. For your hair to truly come clean, put a small amount of baking soda in your hand (about the size of a quarter) along with your favorite shampoo. Lather as usual, then take an extra minute or two to make sure you've rinsed thoroughly. Your hair will feel noticeably cleaner and more manageable.

SEE ALSO Hair Revitalizer (salt) on page 45; Hair-Clarifying Rinse (lemons and salt) on page 92; and Herbal Hair Rinse (vinegar) on page 135.

DANDRUFF FIGHTER

1 handful BAKING SODA

For problem dandruff, especially the kind that accompanies an oily scalp, try using straight baking soda (without shampoo) as another means for cleaner hair. To use, wet hair thoroughly and then work in a small handful of baking soda as you would shampoo, taking care to rub delicately near the scalp. Make sure to rinse thoroughly. While hair may feel dry after the first few washes, it will eventually take on a lovely luster.

SEE ALSO What Else Can It Do? (salt) on page 46 and Dandruff and Oily Hair Fighter (lemons) on page 91.

HOMEMADE DEODORANT

1/2 cup (110g) BAKING SODA
1/2 cup (65g) cornstarch

Most commercial antiperspirants contain chemicals, such as aluminum, that control perspiration by blocking pores so they can't release sweat. If you'd prefer a more natural way to control body odor, consider whipping up a batch of homemade deodorant powder to neutralize odors. Mix the baking soda and cornstarch and store in a shaker-style container. (The amount should last a month or so.) Sprinkle a small amount on a damp washcloth and pat onto underarms (do not rinse afterwards). TIP: Add finely ground dried lavender to add a pleasant, natural scent to the powder.

DIAPER RASH SOOTHER

3 tablespoons (42g) BAKING SODA
Lukewarm water

A baby's tender skin can be ultra-sensitive to uric acid (a product of urine), resulting in an uncomfortable rash that can make wearing a diaper quite unpleasant. To relieve the sting, try bathing your baby in between diaper changes. Prepare a lukewarm bath and mix in 3 tablespoons (42g) baking soda. Afterwards, let the affected area air-dry before diapering (apply a barrier of ointment to afford the skin an opportunity to heal). Repeat as often as possible until the rash clears up.

SEE ALSO Baby Clothes and Cloth Diaper Cleaner (baking soda and vinegar) on page 171.

LOCKER ROOM DEODORIZER

1 16-ounce box (454g) BAKING SODA

If you frequent the gym and maintain a personal locker, the damp conditions from the showers can make the contents of your locker all the more odorous. To keep odors at bay, keep a box of baking soda in the back of your gym locker. And don't forget that baking soda can work double duty for you. If foot odors are a particular problem, for example, try sprinkling a little baking soda into your shoes before placing them in the locker.

FOOT ODOR ELIMINATOR

2 gallons (7.5 liters) warm water

1/2 cup (110g) BAKING SODA

4 drops lavender essential oil (optional)

If embarrassing foot odor has you down, don't hide your feet in more layers of socks; instead, try pampering them with a baking soda soak once a week. Fill a plastic tub with the warm water and baking soda. If you like floral scents, add a few drops of lavender oil, too. Soak your feet for 15 minutes and then pat dry. The baking soda will gently restore your skin's pH levels and neutralize the odor-causing bacteria responsible for the stink.

SEE ALSO Foot Odor Remedy (vinegar) on page 132 and Automatic Shoe Freshener (baking soda) on page 172.

What Else Can It Do?

PREVENT FLATULENCE

Adding a small pinch of baking soda to beans while they're cooking will significantly reduce the beans' aftereffects.

Household Uses

ALL-PURPOSE BAKING SODA CLEANER

1 cup (250ml) warm water
$1/4$ cup (60ml) dishwashing liquid
1 16-ounce box (454g) BAKING SODA

A simple squirt of this all-purpose cleansing scrub and a damp sponge are all you need to clean most non-porous surfaces around the house, like countertops, sinks, tubs, and toilets. Start by putting the warm water in a large mixing bowl and stirring in the dishwashing liquid (stir slowly to avoid creating a lot of bubbles). Next, add the box of baking soda and stir until the mixture forms a smooth, uniform consistency. Transfer the mixture to a jar and use as needed around the house (wait 30 minutes before putting a lid on the fresh batch to let the carbon dioxide bubbles escape). TIP: Be forewarned that the solution will separate between uses, but if you set the jar upside down for a few minutes, the layers will loosen; you can then follow up by giving the contents a gentle stir.

SEE ALSO All-Purpose Lemon Cleaner (lemons and vinegar) on page 72.

RED WINE REMEDY FOR CARPETS

2 or 3 handfuls BAKING SODA

Red wine has a fantastic ability to stain, owing to its acidic nature and abundance of colorful plant compounds. Whatever you do, don't rush to blot a spill for you may inadvertently drive the stain deeper into the pile of the carpet. Instead, cover the area thoroughly with baking soda and let it absorb the liquid, then scoop up the baking soda. Apply a fresh layer of baking soda and let dry overnight. Then simply vacuum in the morning.

SEE ALSO What Else Can It Do? (salt) on page 42.

What Else Can It Do?

CLEAN BATHROOM TILE

If dark mold and mildew are starting to invade your bathroom's grout, you'll likely want to avoid commercial cleansers. Aside from producing noxious fumes, harsh cleaning compounds may not be so gentle on the surrounding tile and tub. Take a more gentle approach by using a baking soda paste on those stains: Add enough water to 3 tablespoons (42g) baking soda to form a paste, then apply to mold and mildew areas with an old toothbrush and scrub until gone. While you're at it, the baking soda paste is also good for hard water or rust stains on the tile. Rinse thoroughly when you're done.

SEE ALSO Mildew Stain Remover (lemons and salt) on page 84 and Plastic Shower Curtain Cleaner (vinegar) on page 136.

What Else Can It Do?

FRESHEN UPHOLSTERY AND CARPETS

If you notice pet, smoke, or cooking odors lingering around the house, chances are they've sunk into your upholstered furniture and carpet. To minimize such smells and freshen your furniture in between professional cleanings, sprinkle baking soda on carpets, pillows, and seat cushions. Let it sit for 30 minutes to neutralize the odors, and then vacuum thoroughly.

DRAIN CLEARER

3 spoonfuls BAKING SODA

After using baking soda in its natural powder form to clean or deodorize, don't just throw it out! Make a habit of pouring a few spoonfuls down the drain, followed by a few minutes of hot water from the faucet. It's a great way to cut down on the grease and grime that naturally accumulates and keep any associated odors to a minimum.

SEE ALSO Clogged Drain Clearer (salt and baking soda) on page 35 and Drain Clearer (vinegar and baking soda) on page 120.

SEPTIC TANK MAINTENANCE

1 cup (220g) BAKING SODA

Flushing 1 cup (220g) baking soda into your septic system once a week is an easy way to maintain a neutral pH range in the tank. This basically means that the bacteria present are able to thrive, and, thus, process the other organic matter. A bit of baking soda also helps keep those rotten egg smells that sometimes accompany a septic system to a minimum.

TOILET BOWL CLEANER

1 cup (250ml) VINEGAR
1/2 cup (110g) BAKING SODA

To give your toilet an all-natural clean, start with vinegar. Pour it into the bowl and let it sit for 1 hour. The mild acidity will jump-start the cleaning process and make the next step much easier. To finish the job, dip the toilet brush in the bowl and then sprinkle with the baking soda. Scrub the inside of the bowl until clean, adding more baking soda to the brush as needed.

STUFFED ANIMAL CLEANER

½ cup (110g) BAKING SODA

Many plush toys are not safe in the washing machine, which makes the problem of cleaning them all the more challenging. To rid stuffed animals of unpleasant odors, place them in a plastic bag with ½ cup (110g) baking soda. Shake the bag gently and let sit overnight. In the morning, brush the baking soda off the toy and it should smell noticeably fresher.

SEE ALSO Baby Toy Disinfectant (vinegar) on page 143.

What Else Can It Do?

REMOVE CRAYON MARKS

Whether your budding artist decorates your walls with some inspired crayon drawings or a crayon accidentally runs through the washing machine, baking soda can save the day. If you're facing homegrown graffiti, try scrubbing the wall with a paste of baking soda and water. In the case of the load of laundry, wash the affected items again in the hottest water possible (according to the garment instructions), but don't add soap; instead, add at least ½ 16-ounce box (230g) baking soda, or up to a full box (454g) for real disasters. Repeat if necessary.

SEE ALSO Pre-Treatment for Kid Stains (baking soda and vinegar) on page 170.

What Else Can It Do?

REMOVE MUSTY ODORS FROM BOOKS

Used booksellers swear by the power of baking soda. Sprinkle a spoonful or two in between the pages of an old book, and any lingering musty odors will soon be neutralized. After a week has passed, open the book and fan the pages to get rid of the powder that remains.

HOLIDAY ORNAMENT CARE

1 16-ounce box (454g) BAKING SODA

Those family keepsakes that we hold so dear—holiday ornaments—are especially vulnerable to developing musty odors during the off-months if they're kept in places susceptible to dampness (e.g., the basement or back of a closet). To minimize the chance of an unpleas-ant unpacking, consider placing a box (454g) of baking soda in the storage container before putting away holiday decorations. If you find that things are still stale, try sealing them in a plastic bag for a few days with 2 tablespoons (28g) baking soda to neutralize the smell.

Outdoor Uses

LAWN FURNITURE CLEANER

¼ cup (55g) BAKING SODA
1 quart (1 liter) warm water

If your old patio furniture is looking a little weather-beaten before the season starts, brighten it up with a good baking soda wash. Mix the baking soda into the warm water and stir to dissolve. Wipe down all surfaces with the solution and rinse clean with fresh water.

SEE ALSO Prevent Wicker Furniture from Yellowing (salt) on page 59.

What Else Can It Do?

NEUTRALIZE BATTERY ACID CORROSION

Changing a leaking car battery can be dangerous if the acid gets on your skin. Protect yourself by applying a thick layer of baking soda to neutralize any spills. Wait until the baking soda stops fizzing and then wipe clean with paper towels. A 16-ounce box (454g) of baking soda will neutralize about 2 cups (500ml) of acid.

Do You Know...

Aside from numerous household uses, there are many industrial applications for baking soda that leave the environment noticeably cleaner. Many cities use it to remove paint and grime from large surface areas, such as buildings that have been marred with graffiti (it's much less toxic than many commercial cleaners). Baking soda also helps reduce lead contamination in drinking water and removes acidic byproducts from smokestack emissions.

OIL SPILL ABSORBER

1 16-ounce box (454g) BAKING SODA
1 16-ounce box (454g) cornmeal

If your car is leaking a little oil, you're most likely to see the evidence on your garage floor or driveway. Fortunately, the cleanup will be easy (and cheap) to tackle with this formula—unlike the leak itself. Mix the baking soda and cornmeal in a large plastic container (the amounts here should be enough for several spills). Sprinkle lightly on a fresh spill to soak it up. Let dry overnight, then sweep or use a shop-vac to get rid of it in the morning.

CAR EMERGENCY CLEAN-UP KIT

1 16-ounce box (454g) BAKING SODA

1 travel-size box baby wipes

In order to be prepared for any kind of mess that can happen in the car, keep a box of baking soda and a small box of baby wipes handy. Sprinkle some baking soda on muddy spots on the floor mats, and they'll be much easier to vacuum up after the mud dries. Or, if you happen to spill a little gas when you're refueling at the pump, sprinkle some baking soda on a baby wipe and use it to clean your fingers. There are too many uses to list them all here, but you'll be happy that you have both items on hand.

BLACK SPOT FUNGUS PROTECTION

2 quarts (2 liters) warm water

$1^1/_2$ teaspoons (7g) BAKING SODA

$^1/_2$ teaspoon (2.5ml) dish soap

Black spot fungus causes very distinct ugly black spots on the leaves of rosebushes—just as the name implies. Left unchecked, the fungus can leave a bush stripped of leaves and may even affect the flowers; and a bush that's severely affected can have a much harder time over the winter months. To keep the fungus at bay, mix together the water, baking soda, and dish soap and spray on leaves once a week in the early morning hours.

GARDEN TOMATO SWEETENER

1 small spoonful BAKING SODA

Whether you grow tomatoes in a container garden or a full-fledged vegetable patch, work a small spoonful of baking soda into the soil when you transplant your seedlings. The resulting tomatoes will be noticeably sweeter at the height of summer because the baking soda lowers their acidity.

What Else Can It Do?

GET RID OF UNWANTED CRABGRASS

Crabgrass, an invasive weed that grows abundantly, presents real problems for homeowners who want to get rid of it without killing the surrounding grass. Fortunately, there's the baking soda solution: To begin, spray the crabgrass area with a fine mist from your garden hose (you want it wet, but not soaking). Next, fill a sock with ½ cup (110g) baking soda and gently hit the side of it with your hand to release a fine layer of dust just above the crabgrass. Within 24 hours you should see that the crabgrass has turned dark, without affecting the surrounding grass.

SEE ALSO **Poison Ivy Killer (salt) on page 58 and What Else Can It Do? (vinegar) on page 148.**

What Else Can It Do?

GROW GORGEOUS FLOWERS

Some types of flowers, especially geraniums, begonias, and hydrangeas, prefer slightly alkaline soils. To evaluate the acidity level of your garden soil, put a spoonful of soil in a small bowl. Add a few drops of water and a pinch of baking soda. If you see any fizzing occurring, the soil's pH level is most likely slightly acidic (less than 5.0). To put your garden soil into the proper pH range, add a tablespoon (14g) of baking soda to a gallon (4 liters) of water and use it to water your plants every other week. You'll be amazed with the results.

POOL TOY MILDEW PREVENTION

1 handful BAKING SODA

Inflatable pool toys, including air mattresses, balls, and inner tubes, are susceptible to mold and mildew once they're deflated and stored in the off-season. To minimize this risk, first make sure to dry them thoroughly in the sun for a couple of hours. And for an extra measure, sprinkle each item with a handful of baking soda before rolling up for long-term storage. The baking soda will help absorb any moisture they may be exposed to until next summer.

BARBECUE FLARE-UP TAMER

2 cups (500ml) water
1 teaspoon (5g) BAKING SODA

A barbecue grill can get out of hand when the fat from cooking meat drips directly on the coals and ignites. And while your immediate instinct might be to douse it with water, doing so can send grease flares higher, as well as seriously hamper your heat source. Instead, keep a spray bottle filled with water and baking soda nearby. Spray lightly when flames shoot up, and the baking soda will fight the fire without completely cooling the coals.

SEE ALSO What Else Can It Do? (salt) on page 59.

What Else Can It Do?

CONTROL ODOR IN THE COMPOST BIN

Compost, known to many gardeners as "black gold" for its amazing ability to enrich the soil and nourish plants, can also produce some amazing odors due to bacterial processes that break down plant material. To help neutralize the odor and lower the acidity level of the pile, try adding a few spoonfuls of baking soda into the mix every other time you turn the heap.

---

1 16-ounce box (454g) BAKING SODA

Don't forget to buy an extra box of baking soda the next time you go camping. Its many uses will come in handy during your trip, whether as a toothpaste or deodorant alternative, a mosquito bite remedy, or a fire extinguisher. And once you're back home, sprinkle what remains in the box on your tent and sleeping bags to remove odors before storing for the next trip.

SEE ALSO Tooth-Whitening Toothpaste on page 174; Bug Bite Remedy on page 175; Homemade Deodorant on page 177; and Barbecue Flare-Up Tamer on page 191.

What Else Can It Do?

DECORATE A CHRISTMAS TREE

Much better for the environment than a snow spray that comes in an aerosol, baking soda can be dusted on a Christmas tree to mimic snow.

Pet Uses

CAT BOX DEODORIZER

1 cup (220g) BAKING SODA

There's no need to buy the more expensive fragrance-filled brands of cat litter when you have some extra baking soda in the house. Fill the litter box halfway with cat litter and then spread a layer of baking soda (about 1 cup or 220g), followed by another layer of litter.

PET BED DEODORIZER

1/2 cup (110g) BAKING SODA

A little baking soda can be a great way to keep your pet's bed smelling fresh between washings. Just sprinkle 1/2 cup (110g) on top and let it sit for a few hours to absorb unpleasant odors. Then take it outside and give it a few good shakes; or, better yet, use a handheld vacuum to get rid of some of the hair as well.

SEE ALSO Pet Bed Cleaner (salt) on page 62.

SKUNKED-DOG CLEANER

1 quart (1 liter) hydrogen peroxide

¼ cup (55g) BAKING SODA

1 teaspoon (5ml) dish soap

There is simply no way to describe the smell of a freshly skunked dog. Fortunately, you can turn that smell into a distant memory with the amazing combination of hydrogen peroxide and baking soda. Mix the peroxide with the baking soda and dish soap in a large plastic bucket (do not store the mixture in a sealed container—it will produce lots of bubbles and explode). Wear latex gloves and bathe your dog outside, if possible. Take care to work the mixture into the fur, concentrating on the areas sprayed, while keeping it away from your dog's face and eyes. Let stand for 10 minutes and then rinse thoroughly with fresh water. See your vet if your dog shows any sign of injury (some skunks are rabid).

What Else Can It Do?

REMOVE "WET DOG" ODOR

There's nothing like the smell of a wet dog, especially during a car ride home after a fun day spent playing catch at the beach. To make the drive home a little more tolerable, make sure to pack a box of baking soda. Before getting in the car, towel dry your dog as much as possible and then lightly sprinkle the fur with a fine layer of baking soda. As the powder dries, your dog will be able to shake it off. In the meantime, the baking soda will do an amazing job of neutralizing the smell.

What Else Can It Do?

FIGHT ODOR IN A HAMSTER CAGE

To help fight the unwanted smells associated with hamsters and other small rodent pets, try washing out their cage with a baking soda and water solution: Remove all pine shavings and toys, and fill a pan with warm water. Add $1/2$ cup (110g) baking soda and a few drops of dish detergent. Use a soft scrub brush to wash the cage and toys with the baking soda mixture, rinse thoroughly with hot water, and dry with a soft cloth. The baking soda will not only loosen any dried matter sticking to the bottom of the cage, but it will also help neutralize odors.

CARPET CLEANER FOR PET VOMIT

2 or 3 handfuls BAKING SODA

A sick pet can lead to all kinds of undesirable stains—from grass to bile to whatever was most recently in the food bowl—so make it a point to have some baking soda on hand for fast action. Start by sprinkling baking soda on the mess to neutralize the acids and absorb some of the liquid. Let the baking soda dry, then scrape or vacuum it up. Follow up with a splash of club soda if any stains remain and blot dry. NOTE: Whatever you do, don't scrub it while it's wet, as this can drive the stain deeper into the fibers.

Kids' Activities

MAKE A SODA BOTTLE "BOAT" SAIL ACROSS THE WATER

4 or 5 generous spoonfuls BAKING SODA

4 or 5 squares toilet paper

1 20-ounce (591ml) plastic bottle (empty)

1/4 cup (60ml) VINEGAR

Here's a fun activity to make a kiddie pool seem even more exciting: Lay 4 or 5 squares of toilet paper on a table and put a generous spoonful of baking soda in the middle of each. Working one sheet at time, gather the four corners together, twist loosely to secure the contents, and gently push the bundle through the mouth of an empty 20-ounce (591ml) plastic bottle. When all of the bundles are in the bottle, top it off with 1/4 cup (60ml) vinegar and give the bottle cap one simple twist so that it is not tightly secured. Lay the bottle on its side in the pool or bathtub, then watch it skim along the surface as the chemical reaction inside sends millions of bubbles shooting out the back (the loosely capped end). TIP: Add marbles to weigh down the bottle opening so that the back end is in the water.

2 teaspoons (9g) BAKING SODA

Tall clear plastic or glass container

Water

VINEGAR

Food coloring

Various small objects (rice, raisins, small pieces of dried pasta, pieces of fruit)

This fun science experiment shows how carbon dioxide helps objects swim. When baking soda (a base) and vinegar (an acid) are combined, a chemical reaction occurs that produces carbon dioxide. To start, almost fill the container with three parts water and one part vinegar. Slowly and carefully add 1 teaspoon baking soda. This is going to react with the vinegar so be ready for some fizzing and a possible overflow. Once the fizzing dies down, slowly and carefully add the second teaspoon of baking soda, and follow up with the food coloring once the bubbles settle. With the added food coloring, you will see the liquid moving. Drop in an object. It will most likely sink, but once enough bubbles attach to it (thanks to the carbon dioxide!), it will swim. Keep experimenting with different items to watch how they move. TIP: If the movement appears to be slowing down, slowly and carefully add another teaspoon of baking soda.

♦ ♦ ♦

Housekeeping Calendar

Many of the natural formulas in this book work best when they're incorporated into a regular cleaning and maintenance cycle. Of course not all the regular chores most people undertake are covered in this book, and every household has its own unique needs, so customize the following tasks into a plan that works for you. The point is to have a plan and stick to it!

HOUSEKEEPING CALENDAR

EVERY DAY

Clean sinks and wipe down countertops

Wash pots and pans

Disinfect cutting boards

WEEKLY

Clean coffeepot

Deodorize the garbage disposal

Deodorize carpets and upholstery

Mop floors

Clean mirrors and glass surfaces

Scrub toilets

Clean bathtubs

Wash linens and towels

MONTHLY

Clean refrigerator

Clean teakettle

Clean bathroom tile

Clean showerhead

Clean toothbrushes and hairbrushes

Clean window blinds

Flush septic system

Deodorize pet beds

Clean windshield wiper blades

Deodorize Humidifier

SALT	LEMONS	VINEGAR	BAKING SODA
page 35	page 72	pages 137, 138	pages 166-167, 180
pages 29-30	pages 76-77, 78	page 121	page 164
page 32	pages 79, 81	page 118	
pages 32-33		page 118	
	page 79		page 182
			page 182
		page 139	
	page 100	page 138	
			page 183
	page 99		page 180
pages 40-41	page 82	page 126	page 169
page 31			page 161
		page 119	
			page 181
		page 137	
		page 134	page 174
		pages 140-141	
			page 183
page 62			page 193
		page 149	
	page 99		

HOUSEKEEPING CALENDAR

IN THE SPRING

Clean windows

Polish and clean woodwork, walls, and blinds

Test garden soil acidity; adjust if necessary

Protect rosebushes from black spot fungus

Pack winter clothes with sachets

IN THE SUMMER

Protect against ants and other pesky bugs

Plant tomatoes

Protect against odors in the compost bin

Protect against weeds

Clean lawn furniture

IN THE FALL

Clean windows

Polish and clean woodwork, walls, and blinds

Clean pool toys and store

Clean lawn furniture and store

IN THE WINTER

Coat car windows to prevent icing

Remove ice from sidewalks

Clean and care for leather shoes

Pack holiday ornaments

SALT	LEMONS	VINEGAR	BAKING SODA
	page 100	page 138	
	page 100	pages 140-141	
			page 190
			page 188
	page 98		page 170
pages 55-56, 57, 62	pages 102, 105-106, 107	pages 146, 149	
			page 189
			page 191
page 58		page 148	page 189
			page 186
	page 100	page 138	
	page 100	pages 140-141	
			page 190
page 59			
page 60			
page 60			
		page 127	
			page 185

Index

About the Author

SHEA ZUKOWSKI is an editor and writer specializing in earth-friendly information and natural living. A former senior editor for Rodale Books, she spent years seeking out the best advice related to cooking, health, nutrition and pets. She lives in the blissfully quiet town of Emmaus, Pennsylvania, with her husband and two sons.